Land Warfare: Brassey's New Battlefield Weapons Systems and Technology Series into the 21st Century
Volume 2

CANNONS

Land Warfare: Brassey's New Battlefield Weapons Systems and Technology Series into the 21st Century

---

Executive Editor: Colonel R G Lee OBE, Former Military Director of Studies, Royal Military College of Science, Shrivenham, UK.

Editor-in-Chief: Professor Frank Hartley, Vice Chancellor, Cranfield University.

The success of the first and second series on Battlefield Weapons Systems and Technology and the pace of advances in military technology have prompted Brassey's to produce a new Land Warfare Series. This series updates subjects covered in the original series and also covers completely new areas. The new books are written for military personnel who wish to advance their professional knowledge. In addition, they are intended to aid anyone who is interested in the design, development and production of military equipment. Series 2 consisted of:

Volume 1	Guided Weapons – R G Lee
Volume 2	Explosives, Propellants and Pyrotechnics – A Bailey and S G Murray
Volume 3	Noise in the Military Environment – R F Powell and M R Forrest
Volume 4	Ammunition for the Land Battle – P R Courtney–Green
Volume 5	Communications and Information Systems for Battlefield Command and Control – M A Rice and A J Sammes
Volume 6	Military Helicopters – E J Everett–Heath, G M Moss, A W Mowat and K E Reid
Volume 7	Fighting Vehicles – Col T W Terry, S Jackson, Col C E S Ryley, B E Jones and P J H Wormell
Volume 8	Surveillance and Target Acquisition Systems – Col A Figgures, Lt–Col M McPherson, Prof A LA Rogers, Dr P S Hall, T K Garland–Collins and J A Gould
Volume 9	Radar – Dr P S Hall
Volume 10	Nuclear Weapons: Principles, Effects and Survivability – Charles C Grace
Volume 11	Powering War – Modern Land Force Logistics – P D Foxton
Volume 12	Command and Control Support Systems in the Gulf War – M A Rice and A J Sammes

Available in Series 3:

Volume 1	Military Ballistics: A Basic Manual – G M Moss, D W Leeming and C L Farrar

CANNONS

D F Allsop
Royal Military College of Science, Shrivenham, UK

Any views expressed are those of the author and do not necessarily represent the views of the Department/HM Government.

Brassey's
London • Washington

First English Edition 1995

UK Editorial offices:
Brassey's Ltd, 33 John Street, London WC1N 2AT
UK Orders:
Marston Book Services, PO Box 269, Abingdon, Oxford
OX 14 4SD
North American Orders:
Brassey's Inc, PO Box 960, Herndon, VA 22070.

Library of Congress Cataloging in Publication Data
available

British Library Cataloguing in Publication Data
A catalogue record for this book is available from the British Library

ISBN 1 85753 104 3 Flexicover

Typeset by Florencetype Ltd, Stoodleigh, Devon
Printed by The Alden Press, Oxford

Preface

This series of books is written for those who wish to improve their knowledge of military weapons and equipment. It is equally relevant to professional soldiers, those involved in developing or producing military weapons or indeed anyone interested in the art of modern warfare.

All the texts are written in a way which assumes no mathematical knowledge and no more technical depth than would be gleaned from schooldays. It is intended that the books should be of particular interest to army officers who are studying for promotional examinations, furthering their knowledge at specialist arms schools or attending command and staff schools.

The authors of the books are members of the staff of the Royal Military College of Science, Shrivenham, which comprises a unique blend of academic and military experts. They are not only leaders in the technology of their subjects, but are aware of what the military practitioner needs to know. It is difficult to imagine any group of persons more fitted to write about the application of technology to the battlefield.

During the next decade, it is probable that the biggest interest area for gun development will be in the field of cannons. They are used in aircraft, on ships, and on land in both mounted and unmounted roles.

There is a need for both improved performance and reduced system size. The achievement of these conflicting requirements will require new ideas and technology. There are positive signs that these are being developed.

This volume describes the present and emerging technologies in this fast development area of military weaponry.

Acknowledgements

The author is grateful for the assistance of the following cannon and ammunition manufacturers for providing information, photographs and drawings without which it would have been impossible to produce this book:

Allient Tech Systems
Bofors Weapon Systems
British Aerospace Defence Royal Ordnance Division
Giat Industries
Marconi
Martin Marietta
McDonnell Douglas
Oerlikon-Contraves
Rheinmetall GmbH
Santa Barbara SA

Thanks are also due to Mrs Jean Mosley for producing all of the drawings and to Cranfield University Royal Military College of Science for permission to use photographs of weapons and ammunition in their collection.

Contents

List of Tables		viii
List of Figures		ix
Chapter 1	Introduction	1
Chapter 2	Self-Powered Cannon	4
Chapter 3	Cannon Using External Power	22
Chapter 4	Barrel Design	36
Chapter 5	Cannon Ammunition Types	47
Chapter 6	Ammunition Components	67
Chapter 7	Recoil Units	77
Chapter 8	Ammunition Handling	88
Chapter 9	Cannon Weapon Systems	96
Chapter 10	Future Cannon Developments	107
Glossary		119
Index		126

List of Tables

Table 1.1	Main types of medium-calibre ammunition	3
Table 10.1	Number of rounds that can be carried in the same storage space in the Bradley M2 and the Warrior IFV	112
Table 10.2	Overall length of a complete round of ammunition for different calibres	112

List of Figures

FIG. 2.1a	The breech block is released and moves forward under the influence of the breech block spring. The breech block feeds a round into the chamber	6
FIG. 2.1b	The cartridge is ignited before the round is fully chambered. The breech block is slowed down by the increase in pressure in the chamber	6
FIG. 2.1c	The bolt is accelerated rearwards by the gas pressure and the empty case is extracted. Energy is stored in the breech block springs for the counter-recoil phase of the cycle	6
FIG. 2.1d	The breech block is stopped by the buffer and counter recoil commences	6
FIG. 2.2	30mm Rheinmetall Mk108 aircraft cannon	7
FIG. 2.3a	With the round chambered, the breech block is locked to the barrel. Upon firing, the breech block and barrel are driven backwards	8
FIG. 2.3b	The breech block and barrel are slowed by the barrel spring and the breech block return spring. Towards the end of the stroke the breech is unlocked from the barrel	8
FIG. 2.3c	The bolt is held to the rear and the barrel moves forward under the influence of the barrel spring. As the barrel moves forward the empty cartridge case is extracted and ejected	8
FIG. 2.3d	The breech block is released, a fresh round is chambered and the breech block is locked to the barrel	9
FIG. 2.4	Sliding wedge breech of the 30mm calibre RARDEN cannon	10
FIG. 2.5	British Army Bofors 40mm calibre L/70 anti-aircraft cannon	11
FIG. 2.6a	The barrel and breech block move into the forward position and a round is fed into the chamber	12
FIG. 2.6b	The round is fired and the barrel and breech block are pushed to the rear by the recoil forces. After a short distance they are unlocked and the barrel is halted. Additional momentum is given to the breech block by an accelerator lever before the barrel stops moving	12

FIG. 2.6c	The breech block moves to the rear of the body of the weapon under its own momentum	13
FIG. 2.6d	The spent case is extracted and ejected, the breech block is halted by the buffer and the barrel is held to the rear	13
FIG. 2.6e	The breech bolt moves forward and chambers a fresh round	13
FIG. 2.7	Acceleration process	14
FIG. 2.8	0.5in Browning heavy machine gun	15
FIG. 2.9	Protruding lug breech locking	15
FIG. 2.10a	The cannon is ready to fire, the bolt is retained by the sear and the return spring is compressed. A cartridge is ready to be fed into the chamber	16
FIG. 2.10b	The trigger is released and the bolt feeds a cartridge into the chamber. The breech bolt locks to the barrel, the firing pin strikes the primer and the cartridge is fired	16
FIG. 2.10c	After the projectile passes the gas port, the gas piston unlocks the breech block from the barrel. Residual gas pressure accelerates the breech block after the bolt has unlocked. The empty cartridge case is extracted and ejected	16
FIG. 2.10d	A fresh cartridge is chambered when the bolt is pushed forward by the breech block return spring	17
FIG. 2.11	Locking mechanism for pivoting lock bolt	18
FIG. 2.12	Typical revolver cannon	19
FIG. 2.13	Time displacement diagram for a typical revolver cannon	20
FIG. 3.1	Powered Gatling cannon rotor assembly	23
FIG. 3.2	Gatling cannon rotating bolt	24
FIG. 3.3	Operating cycle for a typical Gatling cannon for one barrel	25
FIG. 3.4	GU-8 30mm calibre Gatling cannon (Courtesy: Martin Marietta)	26
FIG. 3.5	Firing cycle for a salvo firing cannon	28
FIG. 3.6	Meroka salvo-firing cannon (Courtesy: Santa Barbara SA)	28
FIG. 3.7	Operating principle of the chain gun	30
FIG. 3.8	Chain gun upper feed operation	31
FIG. 3.9	Chain gun lower feed operation	32
FIG. 3.10	30mm M230 chain gun (Courtesy: McDonnell Douglas)	32
FIG. 3.11	Rheinmetall Rh503 gun mechanism (Courtesy: Rheinmetall GmbH)	33
FIG. 3.12	Dual feed mechanism of the Rheinmetall Rh503 cannon (Courtesy: Rheinmetall GmbH)	34
FIG. 4.1	Diagram of groove profile	37
FIG. 4.2	Forces acting on the body of the projectile	38

FIG. 4.3 Driving band force plotted against shot travel for different rifling twist rates 39
FIG. 4.4 Pressure/time curve for the 30mm calibre RARDEN cannon 41
FIG. 4.5 Typical temperature/time curve at the surface of a 30mm calibre cannon 43
FIG. 5.1a Storage safety. Shutter (1) is locked with the detonator (2) out of line with the firing pin (3) by the security pin (5) and safety lock (6). The firing pin is secured by the split collet (8) 48
FIG. 5.1b Fuze arming. Centrifugal force pulls back the firing pin (3) by the spheres (9) rolling up the inclined face. The split collet (8) and the safety ring (6) open. The security pin (5) is removed allowing the shutter (1) to rotate. The firing pin is only restrained by the spheres (9) 48
FIG. 5.1c Functioning upon impact. On impact, shock is transmitted to the spheres (9) driving them into their recess and releasing the firing pin (3) to strike the detonator (2) 49
FIG. 5.1d Self-destruction. As the spin rate decays the centrifugal force acting on the spheres (9) diminishes. They are driven back into their recess and the firing pin is released 49
FIG. 5.2 Typical 30mm practice projectile (Courtesy: Oerlikon-Contraves) 50
FIG. 5.3 Cross-section through a typical 30mm HE projectile (Courtesy: Oerlikon-Contraves) 50
FIG. 5.4 Typical weights of HE in cannon projectiles 51
FIG. 5.5 Section through a typical 30mm calibre SAPHE round (Courtesy: Oerlikon-Contraves) 52
FIG. 5.6 Cross-section through a typical 30mm calibre APDS round (Courtesy: Oerlikon-Contraves) 52
FIG. 5.7 Cross-section through a typical 30mm calibre APHC round (Courtesy: Oerlikon-Contraves) 53
FIG. 5.8 Bofors 40mm calibre fin-stabilised round and sabot (Courtesy: Bofors Weapon Systems) 54
FIG. 5.9 Typical trajectory diagram for a 25mm calibre HE shell (Courtesy: Oerlikon-Contraves) 55
FIG. 5.10 Typical ranges for HE shells of different calibres 56
FIG. 5.11 Penetration performance of a typical SAPHE round 57
FIG. 5.12 Safety trace for the Royal Ordnance 30mm calibre APDS round and RRTR (Courtesy: Royal Ordnance) 57
FIG. 5.13 Cross-section through tube reduced range training round (Courtesy: Royal Ordnance) 58
FIG. 5.14 Diagrammatic flow around and through the pipe round at high Mach numbers (Courtesy: Royal Ordnance) 58

FIG. 5.15	Diagrammatic air flow for choked tube (Courtesy: Royal Ordnance).	59
FIG. 5.16	Graph showing change in velocity with range for the RRTR (Courtesy: Royal Ordnance)	60
FIG. 5.17	Typical trajectory diagram for a 25mm APDS projectile (Courtesy: Oerlikon-Contraves)	61
FIG. 5.18	Oerlikon AHEAD-MUNITION anti-aircraft shell concept (Courtesy: Oerlikon-Contraves)	62
FIG. 5.19	Details of the Bofors 40mm proximity-fuzed shell (Courtesy: Bofors Weapon Systems)	62
FIG. 5.20	Triggering distances against different targets for the Bofors 40mm proximity-fuzed shell (Courtesy: Bofors Weapon Systems)	63
FIG. 5.21	Bofors 40mm calibre 3P multi-purpose round (Courtesy: Bofors Weapon Systems)	64
FIG. 5.22	Bofors 3P multi-purpose round used to give an air burst (Courtesy: Bofors Weapon Systems)	65
FIG. 5.23	Sequence of events in programming the Bofors multi-purpose round (Courtesy: Bofors Weapon Systems)	65
FIG. 6.1	Different cartridge case rims used with medium-calibre ammunition	68
FIG. 6.2	Typical stress/strain relationship for steel and brass	69
FIG. 6.3	Typical section through the chamber and cartridge case when firing	70
FIG. 6.4	Residual hoop stress in steel and brass cartridge cases	71
FIG. 6.5	Stages in the production of cartridge cases	73
FIG. 7.1	Diagrammatic representation of a gun firing a projectile	78
FIG. 7.2	Force/displacement curve for a typical spring recoil unit	80
FIG. 7.3	Section through one pair of annular ring springs	81
FIG. 7.4	Friction brake with constant braking force	82
FIG. 7.5	Friction brake with progressive breaking force	82
FIG. 7.6	Section through a typical hydraulic unit	83
FIG. 7.7	Recoil diagram for gun mounting with fixed zero position	84
FIG. 7.8	Recoil diagram for a floating mount	85
FIG. 7.9	Twin-baffle muzzle brake	86
FIG. 8.1	Bofors 40mm calibre L/70 cannon fitted to the CV90- (Courtesy: Bofors Weapon Systems)	89
FIG. 8.2	Oerlikon 35mm calibre KDE cannon (Courtesy: Oerlikon-Contraves)	89
FIG. 8.3	Oerlikon 20mm calibre KAB cannon with detachable magazine (Courtesy: Oerlikon-Contraves)	90
FIG. 8.4	Three-round clip ready to be loaded into the RARDEN cannon	91
FIG. 8.5	Flexible feed ducting used with the M230 chain gun (Courtesy: McDonnell Douglas)	93
FIG. 8.6	Linkless conveyor system used with Gatling cannon	94

Fig. 9.1	Martin Marietta AV-8B 25mm calibre gun system (Courtesy: Martin Marietta)	97
Fig. 9.2	Martin Marietta 20mm calibre Vulcan cannon fitted to the F16 (Courtesy: Martin Marietta)	97
Fig. 9.3	Giat 20mm calibre cannon externally mounted on a Gazelle helicopter (Courtesy: Giat Industries)	98
Fig. 9.4	30mm McDonnell Douglas M230 chain gun mounted on the AH-64 Apache helicopter (Courtesy: McDonnell Douglas)	99
Fig. 9.5	Turret and cannon system fitted to the Cobra attack helicopter (Courtesy: Martin Marietta)	100
Fig. 9.6	Bofors Trinity ship-borne anti-aircraft system (Courtesy: Bofors Weapon System)	101
Fig. 9.7	Marconi Marksman anti-aircraft System (Courtesy: Marconi)	102
Fig. 9.8	Goalkeeper anti-missile cannon system (Courtesy: Martin Marietta)	103
Fig. 10.1	Rheinmetall Rh503 cannon with 35mm and 50mm calibre barrels (Courtesy: Rheinmetall GmbH)	108
Fig. 10.2	35mm and 50mm calibre APFSDS-T rounds (Courtesy: Rheinmetall GmbH)	109
Fig. 10.3	25mm calibre Bushmaster and 30mm calibre Bushmaster II cannon (Courtesy: McDonnell Douglas)	110
Fig. 10.4	Section through typical cased telescoped rounds	111
Fig. 10.5	45mm calibre CTA COMVAT cannon (Courtesy: Allient Tech Systems)	113
Fig. 10.6a	For loading, the chamber rotates through 90° to the bore	114
Fig. 10.6b	A round is rammed	114
Fig. 10.6c	The chamber rotates to align with the bore	114
Fig. 10.6d	The round is fired and the gun recoils within the box receiver. Energy is stored in a spring system which returns the chamber to the load position	115
Fig. 10.6e	The chamber is rotated through 90°	115
Fig. 10.6f	A fresh round is rammed, pushing the spent case out of the forward end of the chamber	115
Fig. 10.7	Examples of different feed systems being investigated by Giat (Courtesy: Giat Industries)	116
Fig. 10.8	20mm calibre CTA round and conventional 20mm round (Courtesy: Martin Marietta)	117

1.

Introduction

Definition of a Cannon

There is no internationally agreed definition of a medium-calibre cannon, although it is generally accepted that it is a quick-firing weapon used in the direct-fire role. It is the calibre range which is difficult to define. 20mm is usually accepted as the minimum calibre: smaller calibre weapons than this are normally considered to be heavy machine guns. The upper limit is usually taken to be 50mm calibre. However, the ongoing trend is for these types of weapon to increase in size so that it is prudent not to be too dogmatic about the upper calibre limit. Weapon use is probably a better guide to the definition of a medium-calibre cannon. They are usually deployed mounted on vehicles and used as anti-vehicle and anti-*matérial* weapons rather than anti-personnel weapons. However, even this description becomes blurred where modern, electronically-timed, fuzed fragmenting munitions, which have been developed for use against troops in the open, are concerned.

Target Arrays for Cannon

True lightweight quick-firing cannon were first developed more than 60 years ago for use in aircraft. They replaced the multiple medium- and heavy-calibre machine guns, which were in general use, to increase the terminal effects against both aircraft and ground targets. By increasing the calibre of the weapon it was possible to fill the projectile with high explosive (HE) and to fit a fuze for detonation upon impact. The HE produced high-velocity fragments from the projectile body which considerably increased the kill probability of the weapon. At about the same time, fast-firing cannon were being developed for air defence purposes against low flying aircraft for both land and sea-borne use. Once again the large capacity of the projectile for carrying HE increased the kill probability against a fast-moving target that was difficult to hit.

In the 1960s large scale mechanisation of the infantry began, which resulted in the widespread use of armoured personnel carriers (APCs). These were designed to protect troops from small arms fire so that a new need was created for cannon for use against these newly developed armoured vehicles. The most obvious platform on which to mount cannon was the APC itself. For use against hard targets these weapons required a different type of ammunition from the HE fragmenting round with an impact fuze: a round to defeat armour was required which relied upon the kinetic energy of the projectile to penetrate the target. Anti-aircraft cannon proved to be very effective in the anti-armour role because of their high muzzle velocities. The result is that many medium-calibre cannon have a common ancestry based upon early anti-aircraft weapons.

The same target arrays are still the primary targets of modern fast-firing cannon. The APC has continued to develop into the infantry fighting vehicle (IFV) with even more effective armour protection. Cannon systems have also been developed for defence against missiles and helicopters as well as fixed wing aircraft. The protection level of all target arrays has steadily increased since the first introduction of quick-firing cannon and this has been reflected in the steady increase in the calibres of cannons and in the development of special anti-armour projectiles. These projectiles are akin to tank gun rounds and have considerably increased the armour penetration performance of cannon ammunition. Highly sophisticated, electronic fuzes have also been developed to increase the hit chance of these weapons against agile and fleeting aerial targets.

Cannon Calibres

There is a bewildering variety of cannon sizes used by different nations with, until recently, very little standardisation on calibre and cartridge case configuration. Thus two rounds of nominally the same calibre may be very different. An example of this is the 30mm ADEN/DEFA round that was developed for an aircraft cannon and the 30mm RARDEN, developed originally as an anti-aircraft round for the Oerlikon KCB cannon and later as an anti-armour round for the RARDEN. The 30mm ADEN round has a parallel case of 113mm length, whereas the 30mm RARDEN has a bottle-neck case of 170mm length. The result is that the projectile of the ADEN weighs 245gm and leaves the muzzle at 820m/s, whereas the RARDEN round has a projectile of 360gm and leaves the muzzle at 1070m/s. This results in the muzzle energy of the RARDEN round being approximately twice that of the ADEN round. A better understanding of the potential power (projectile weight and velocity) of a round is therefore given if both the calibre and the case length are considered.

Table 1 shows details of the main types of ammunition for medium-calibre weapons. This list is by no means complete. It contains only those calibres and weapons which are still used extensively today and does not include Soviet calibres. The list could, for example, contain a further five 20mm calibres with the assorted cannon that fire them.

Aircraft cannon have continued to develop and are now one of the most sophisticated weapons systems to be mounted on a vehicle. They have very high rates of fire with complex ammunition handling systems to support the very high feed rate requirement. Attack helicopters are among the latest weapon systems to be introduced to the battlefield. A new role for medium-calibre cannon is as one of the main armaments for these highly effective weapons platforms.

Medium-calibre cannon are highly complex weapons that are often required to achieve high rates of fire, high muzzle velocities and very high levels of accuracy. Their calibre must be sufficiently large to enable sophisticated types of projectile to be used and to achieve the required terminal effects against well-protected

TABLE 1.1
Main Types of Medium-Calibre Ammunition

Designation	Origin	Weapons/Acronyms
20 × 102	USA	Vulcan, VADS PIVADS, CIWS, USM39, M61A1, GAU-4, M168, M197[USA], M621, [F]
20 × 128	CH	Oerlikon KAA, Oerlikon KAB [ex-STG] [CH], Meroka [S]
20 × 139	CH	HS820, Oerlikon KAD [CH], M693 [F], Mauser Mk20, Rheinmetall Rh202 [FRG]
25 × 137	CH	Oerlikon KBA [CH], Mauser Mk25 Mod E, Rheinmetall Rh205 [FRG], ADEN 25 [UK], M811 [F], GAU-12/U, M242 Bushmaster, GE 225 [USA]
25 × 184	CH	Oerlikon KBB [CH]
27 × 145B	FRG	Mauser BK27 [FRG]
30 × 113B	UK/F	ADEN/DEFA-DEFA 551, 552, 553, 554[F], ADEN 30 [UK]. M230 [USA]
30 × 150B	F	Giat Mod.781, Mod 791B [F]
30 × 170	CH/UK	RARDEN [UK], HS831L, Oerlikon KCB [CH]
30 × 173	CH	Oerlikon KCA [CH], Mauser Mk30 Mod.F [FRG], GAU-8/A, GAU-9/A, GAU-13/A [USA]
35 × 228	CH	Oerlikon KDA, KDB, KDC, KDE, KDF [CH], Mauser Mk 35 Mod.G [FRG], Ares TALON 35 [USA]
40 × 311R	SW	40/L60 Bofors [SW]
40 × 365R	SW	40/L70 Bofors [SW]

B = BELTED
R = RIMMED

targets. Additionally, because these weapons are usually vehicle-mounted, they must be light and compact. These are very stringent requirements which can only be achieved with innovative, high quality design. The complexity of these weapons is such that they are expensive to design, build and develop. Their production is therefore confined to a small number of manufacturers that have built up the necessary expertise.

2.

Self-Powered Cannon

Advantages and Disadvantages

The ultimate source of energy for driving self-powered cannon is the burning propellant from the cartridge. It may be used directly or indirectly from the high-pressure propellant gases, or after it has been transferred through the recoil system of the weapon, or through a combination of both. Less than 1 per cent of the muzzle energy is required to cycle the working parts so that the effect upon the weapon's projectile performance is negligible.

However it is used, the energy is only available for the very short time that the propellant is burning; but energy must be available to complete the operating cycle. Self-powered cannon therefore require some form of energy-storing device, such as springs or moving masses, which can be charged in a short time and then return the energy when needed in the course of the firing cycle. Substantial forces arise inside the cannon due to the short energy-transmission times resulting from this mode of operation. Thus cannon of this type tend to be large and heavy so that they are able to withstand these forces.

Although self-powered cannon do not need a power supply to operate them, an external power source is needed for the initial cocking and feeding operation. This may be difficult and time-consuming to achieve by hand, especially in the case of remotely-sited weapon systems such as aircraft cannon, so it is common to fit an external power source for this initial part of the operating cycle. Additionally, it is usual to arrange for the mechanism to stop in the cocked position when the last round from the magazine or belt has been fired to save the effort of re-cocking the weapon after reloading.

Because the operating cycle is powered by the ammunition propellant, the operating reliability is highly dependent upon the reliability and consistency of the ammunition. If a round fails to fire, the weapon ceases to operate and clearing the round is often difficult and time-consuming. However, double feeding of ammunition is not possible during a normal cycle and self-powered weapons have a built-in hang-fire safety because they will not operate until the round fires.

Self-powered cannon are complex; they are also difficult and expensive to develop into reliable and safe systems. Even those of relatively modern design can trace their design ancestry back to the Second World War or before. Those which use external power tend to be of recent design.

Blow-Back Cannon

In blow-back cannon the breech block is at no time rigidly locked to the barrel, so that the force exerted by the propellant gases on the inner base of the cartridge case, and hence to the breech block, is only resisted by the inertia of the breech block. The mass of the breech block must be sufficiently large to ensure that the case remains supported by the chamber until the gas pressure has dropped to a level low enough not to rupture the cartridge case.

Pure Blow-Back

Pure blow-back operation is rarely used in anything other than low-pressure, small-calibre weapons because the weight of the breech block to resist the initial peak gas pressure is usually too large for any other type of weapon.

Delayed Blow-Back

To reduce the weight of the breech bolt, delayed blow-back operation can be used. In this system the breech bolt is made in two parts. As the two parts are pushed to the rear a leverage mechanism acts between them. This accelerates the rearward part and slows the forward part, thus giving the same effect as a heavier breech block. Delayed blow-back is frequently used on small-calibre weapons but rarely on medium-calibre cannon.

Advanced Primer Ignition (API)

The weight of the bolt can also be considerably reduced by using advanced primer ignition. In this type of action the weapon must operate from the open bolt position; the chamber of the weapon is empty and the breech block held to the rear. When the weapon is fired the bolt is released from its position at the rear of the receiver; it picks up a round of ammunition from the magazine or belt and starts to chamber it. Before the round is fully chambered the primer is fired, either mechanically or electrically, with the round and the breech block still moving forward.

Pressure from the propellant gases rises to a peak, at which time the round is fully chambered so that the cartridge case is fully supported by the chamber. The chamber pressure is thus falling as the breech block is pushed to the rear of the receiver. To ensure maximum support to the case, the advanced primer ignition system uses parallel cartridge cases. Propellant gas pressures must be kept fairly low to prevent the cartridge case from rupturing when it is extracted.

The different phases of the operating cycle are shown in Figures 2.1a to 2.1d. The advanced primer ignition blow-back system relies on the primer being ignited at the correct moment when the breech block is moving forward. If primer ignition occurs significantly earlier or later than designed it may pose a considerable hazard to the weapon and the user. Early firing may occur if there is some form of obstruction in the chamber; a common obstruction is the build-up of carbon from the propellant gases. It may cause the case to rupture when the build-up

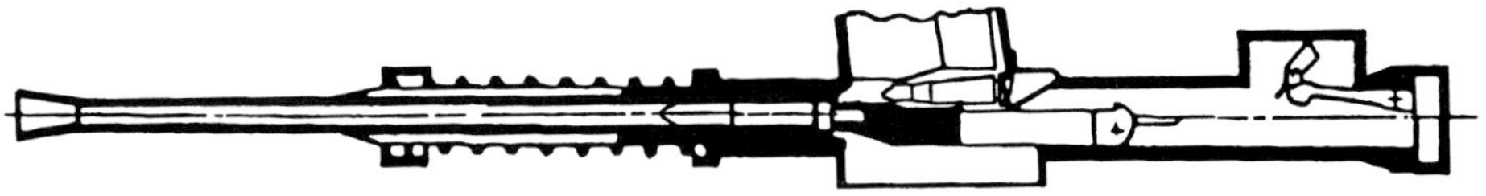

FIG. 2.1a The breech block is released and moves forward under the influence of the breech block spring. The breech block feeds a round into the chamber

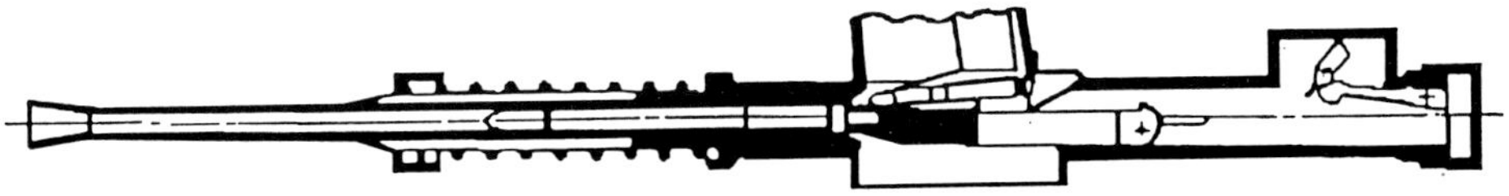

FIG. 2.1b The cartridge is ignited before the round is fully chambered. The breech block is slowed down by the increase in pressure in the chamber

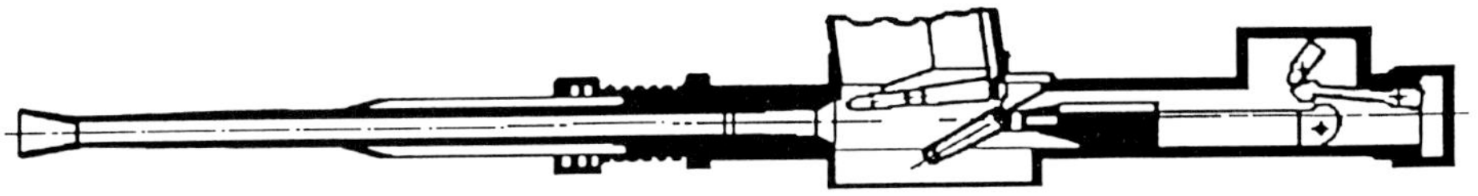

FIG. 2.1c The bolt is accelerated rearwards by the gas pressure and the empty case is extracted. Energy is stored in the breech block springs for the counter-recoil phase of the cycle

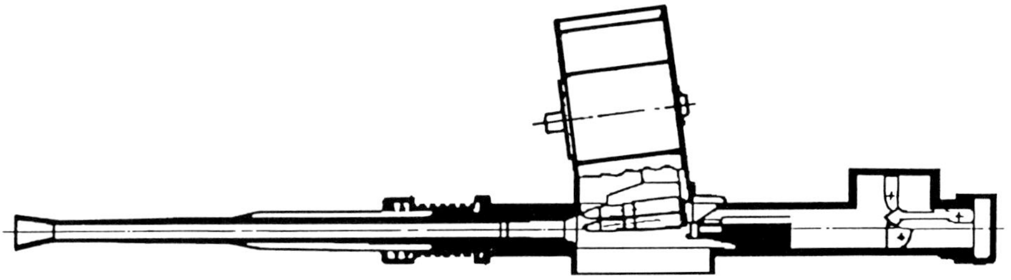

FIG. 2.1d The breech block is stopped by the buffer and counter recoil commences

in pressure becomes excessive before the case is supported by the chamber. Late firing, caused by hang fires, poses similar problems. In this case the chamber pressure does not have time to drop to a sufficiently low level before the support of the chamber is removed. Late firing may also occur if the round is hand fed, because it is fully chambered when the primer is fired.

This form of blow-back operation relies for its functioning upon the momentum imparted to the breech block by the ammunition. It is very susceptible to changes from the different levels of momentum imparted by a variety of ammunition types giving a spectrum of rates of fire. A greater problem may result from firing a mix of different types of ammunition: it results in fluctuations in the firing rate. The consequent high forces in the weapon, especially in the recoil mechanism, can cause premature failure of major weapon components.

Another potential source of damage is the closing of the breech on an empty chamber. The breech block is fairly massive and its return springs are very powerful. Without the cushioning effect which a fired round gives, the impact of the breech block is considerable. A hold-open device is therefore usually fitted to hold the breech block to the rear when the last round in a belt has been fired. This also precludes the need to cock the gun against the breech block springs.

Early aircraft cannon used this type of operating cycle. Figure 2.2 shows the 30mm Rheinmetall Mk108 aircraft cannon developed during the Second World War. The weapon was belt fed and had an initial rate of fire of 600rounds per minute. Attempts were made to double the rate but the highest that was achieved was 900 rounds per minute. Muzzle velocity was a modest 600m/s due to the limitation of low chamber pressures imposed by the blow-back method of operation. This particular weapon is of considerable significance because it was one of the first large-calibre, high rate of fire, ultra-light weight weapons to be fitted into an aircraft. The projectile was filled with high explosive (HE) and was fitted with an impact fuze sufficiently sensitive to operate when impacting the thin skin of an aircraft. Advanced primer ignition can also be achieved by using electrical primers: the electrical circuit is made as the breech block moves forward and chambers the round.

FIG. 2.2 30mm Rheinmetall Mk108 aircraft cannon

The advantages of blow-back cannon are:

- Simple robust design.
- No complicated locking components.
- Reliability.

The disadvantages of blow-back cannon are:

- Heavy breech block or bolt.
- Relatively low rate of fire.
- Shoulderless parallel cartridge case required.
- Only suitable for low powered ammunition.
- Prone to fouling from the residual propellant gases.

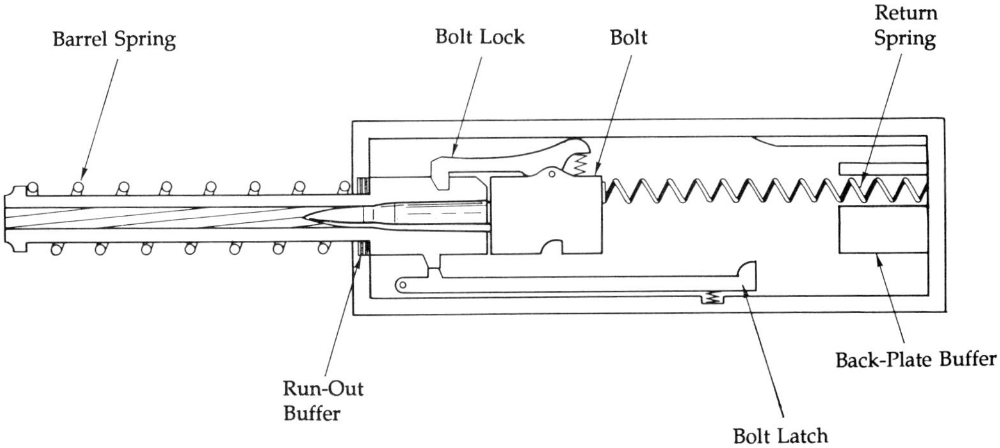

FIG. 2.3a With the round chambered, the breech block is locked to the barrel. Upon firing the breech block and barrel are driven backwards

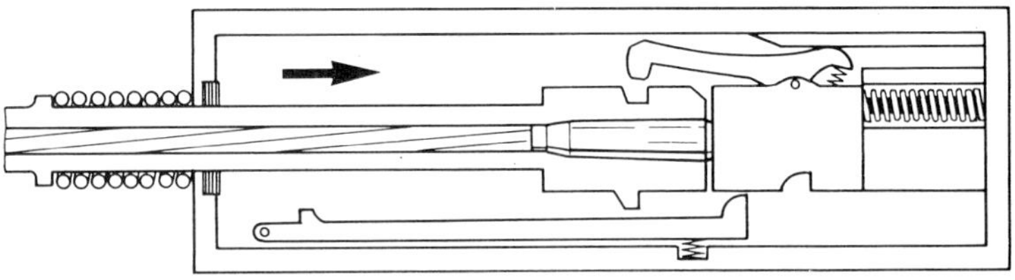

FIG. 2.3b The breech block and barrel are slowed by the barrel spring and the breech block return spring. Towards the end of the stroke the breech is unlocked from the barrel

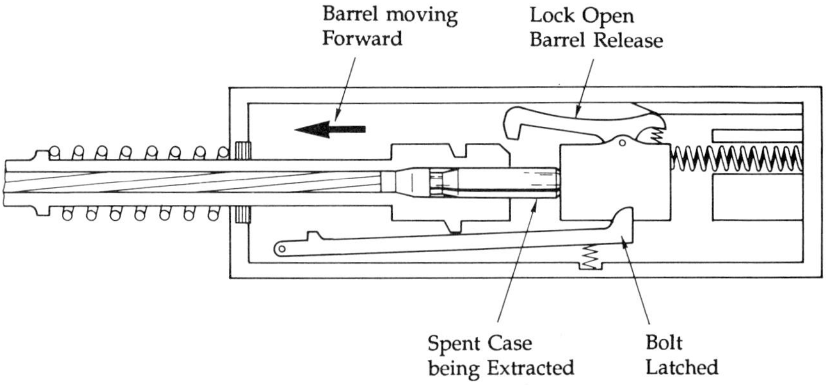

FIG. 2.3c The bolt is held to the rear and the barrel moves forward under the influence of the barrel spring. As the barrel moves forward the empty cartridge case is extracted and ejected

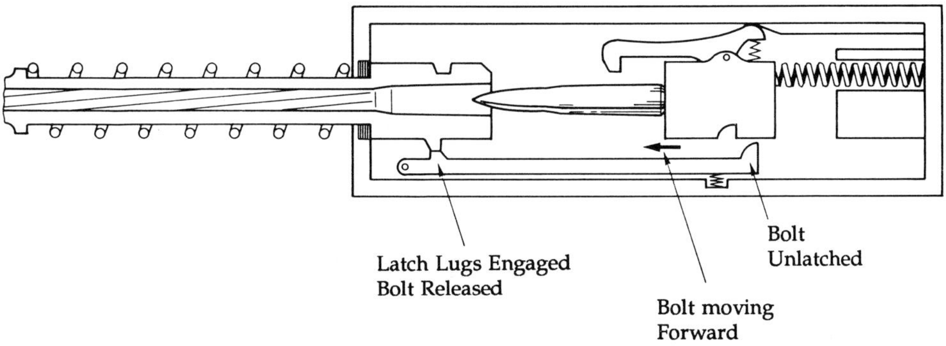

FIG. 2.3d The breech block is released, a fresh round is chambered and the breech block is locked to the barrel

Recoil Operation

As its name implies, recoil operation uses the recoil forces generated when the weapon is fired; it is similar to the blow-back action but in this case both the barrel and the breech block, locked together, are pushed to the rear of the weapon, rather than the breech block alone. This ensures that the cartridge case is fully supported by the chamber all the time that the pressure is high in the cartridge case. Higher chamber pressures can therefore be used to give high muzzle velocities and cartridge cases are not restricted to parallel-sided geometry. Recoil operation can be sub-divided into long-recoil and short-recoil operation.

Long-Recoil Cannon

Long recoil is a system in which the breech block and the barrel are locked together and both move to the rear of the weapon over a distance which exceeds the length of an unfired round. This movement is the source of energy to operate the system: the energy is stored in springs which are compressed by this rearward movement. When the rearward movement is completed, the bolt is unlocked from the barrel and is held to the rear whilst the barrel runs out leaving the empty case on the face of the breech block from where it is ejected. When the barrel has moved all the way forward there is sufficient space for a round to be moved into position for feeding into the chamber.

The recoil forces developed by medium-calibre cannon may be very high with high levels of momentum imparted to the recoiling masses. However, because the recoiling masses are high, their recoil velocities are relatively low. Upon firing, both barrel and breech block recoil at least the length of an unfired round so that the decelerating forces are relatively low. This is a particular feature of the long-recoil system.

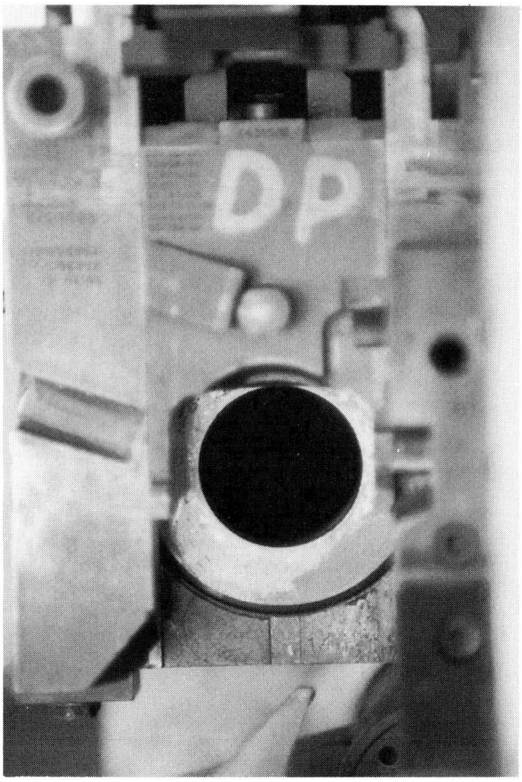

FIG. 2.4 Sliding wedge breech of the 30mm calibre RARDEN cannon

The breech block does not become unlocked until the full recoil stroke has been accomplished, so half the operating cycle elapses before extraction, ejection and feed can commence. Consequently a low rate of fire is associated with this method of operation. However, the long time between the weapon being fired and the case being extracted results in very low residual propellant pressures in the case when it is extracted. Propellant gases are toxic with high levels of carbon monoxide, which can cause problems with turret-mounted weapons. The lower the case pressure at extraction then the smaller is the problem: so long-recoil cannon offer a major advantage for turret-mounted weapons.

The general sequence of events for long-recoil-operated weapons is shown in Figures 2.3a to 2.3d.

Most long-recoil cannon use a sliding wedge breech to minimise the length of the receiver, which is already fairly long, because the recoil distance of the barrel and the breech block is considerable. This type of locking system imposes high transverse forces on the opening mechanism because of the high acceleration forces required to open the breech in the very short time at the end of the rearward movement of breech block and barrel. Figure 2.4 shows the sliding wedge breech for the 30mm calibre RARDEN cannon, which uses the long-recoil operating cycle.

FIG. 2.5 British Army Bofors 40mm calibre L/70 anti-aircraft cannon

During the long-recoil operating cycle the barrel is supported in bearings during its movement backwards and forwards in the receiver of the weapon. These bearings must be well maintained in order to ensure that the accuracy of the weapon is retained. At first sight, this necessity to support the barrel in bearings, which must have an appropriate amount of clearance to work, would appear to have a detrimental effect upon weapon accuracy. However, long-recoil weapons have a reputation for low levels of dispersion, despite the need for barrel bearings.

One of the earliest and best-known anti-aircraft cannon, the Bofors 40mm calibre L/70 cannon, uses the long-recoil operating cycle. Figure 2.5 shows a British Army Bofors anti-aircraft cannon; the rate of fire was 200 rounds per minute. It was magazine fed and the muzzle velocity was 1005m/s for the HE round. Although this weapon is no longer in service with the British Army, similar weapons are still used extensively throughout the world.

The advantages of long-recoil-operated cannon are:

- Positive locking of barrel and breech block.
- Low recoil forces.
- Lightweight design.
- Low gas pressures when the breech is opened.

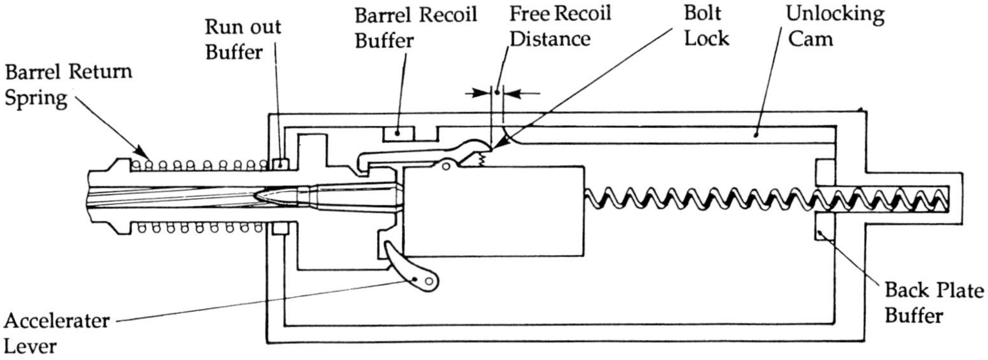

FIG. 2.6a The barrel and breech block move into the forward position and a round is fed into the chamber

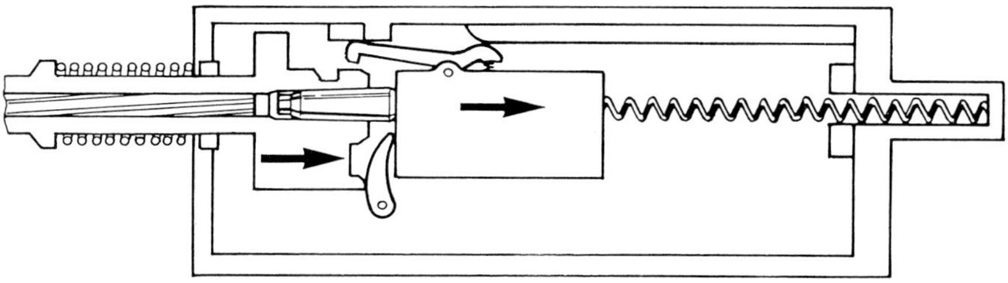

FIG. 2.6b The round is fired and the barrel and breech block are pushed to the rear by the recoil forces. After a short distance they are unlocked and the barrel is halted. Additional momentum is given to the breech block by an accelerator lever before the barrel stops moving

The disadvantages of long-recoil-operated cannon are:

- Long receiver.
- Restricted to low rates of fire.
- Large and expensive barrel bearings.
- Large changes in centre of gravity due to the complete barrel and breech block moving a large distance when firing.

Short-Recoil Cannon

The sequence of operations for the short-recoil cycle are shown in Figures 2.6a to 2.6e

For most short-recoil-operated weapons a lever is used to accelerate the breech block to the rear of the receiver of the weapon. This allows a lightweight breech block to be used. The lever is located between the barrel and the breech block

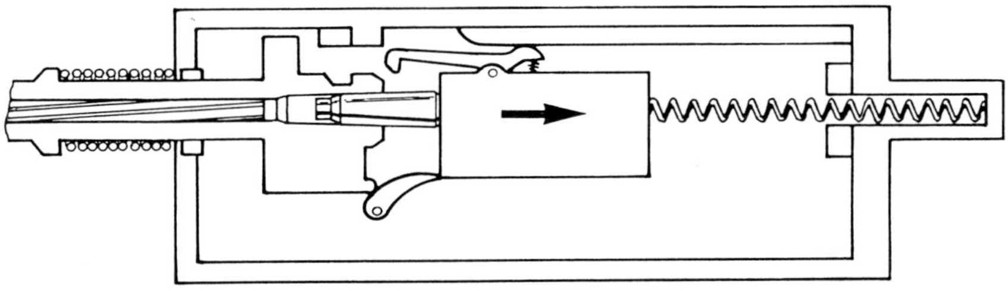

FIG. 2.6c The breech block moves to the rear of the body of the weapon under its own momentum

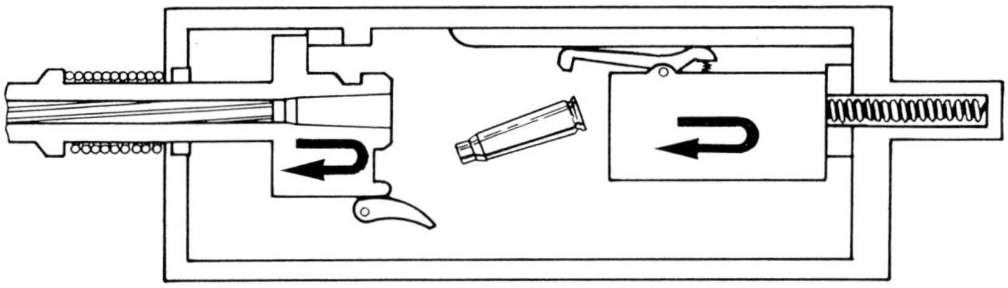

FIG. 2.6d The spent case is extracted and ejected, the breech block is halted by the buffer and the barrel is held to the rear

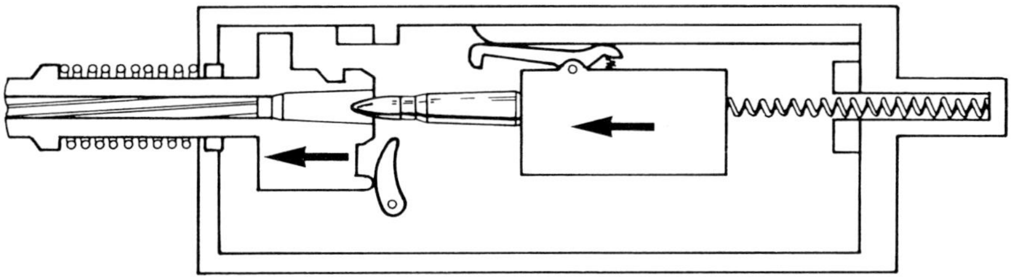

FIG. 2.6e The breech bolt moves forward and chambers a fresh round

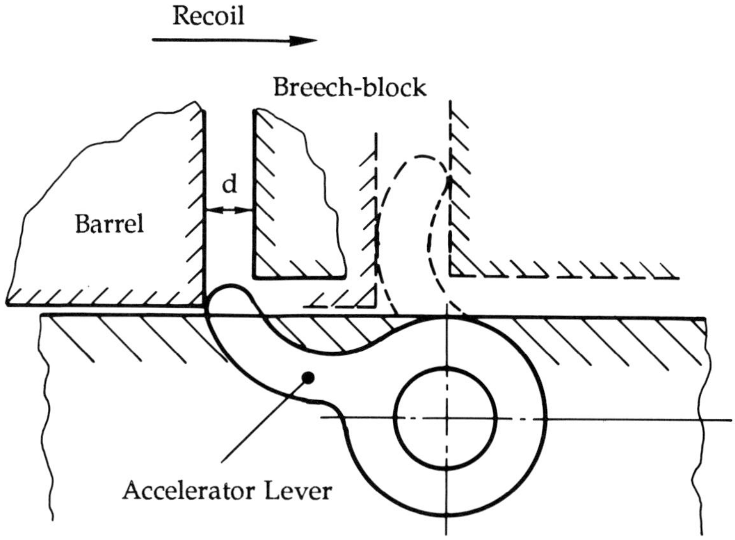

FIG. 2.7 Acceleration process

and is attached to the body of the weapon. As the barrel moves to the rear, the lever is rotated and accelerates the mass of the breech block by virtue of its shape. This action is shown in Figure 2.7.

The short-recoil system was developed to overcome the slow cycle times associated with long-recoil systems. Shorter cycle times are achieved by unlocking the barrel from the breech block after only a short travel of the recoiling masses, typically one-quarter to one-third of the length of the unfired round. After the barrel and breech block have unlocked, the barrel is halted in its rearward motion, but the breech block continues to the rear of the weapon, extracting and ejecting the empty case on the way. It is the action of extracting the case during the rearward motion of the breech block, rather than during the forward motion of the barrel, as in the long-recoil system, which increases the speed of the operating cycle. The breech block must still recoil sufficiently far for there to be enough room to feed a fresh round into the chamber. When the breech block reaches the rear of the weapon it is immediately returned by the breech bolt spring, chambering a fresh round as it does so. The short movement of the barrel and the increased velocity of the breech block caused by the use of an accelerating lever combine to increase the recoil forces exerted by the recoiling masses compared with those of the long-recoil system.

Cannon up to 30mm calibre have been developed using this method of operation. The most successful weapon using the short-recoil operation is the 0.5in. Browning heavy machine gun. The short-recoil system seems ideally suited to this particular calibre. It is possible for the weapon to fire from either the closed or open bolt, although the latter is more usual to reduce the possibility of cook-off from a chambered round left in a hot chamber. Figure 2.8 shows the 0.5in. Browning heavy machine gun mounted on its tripod.

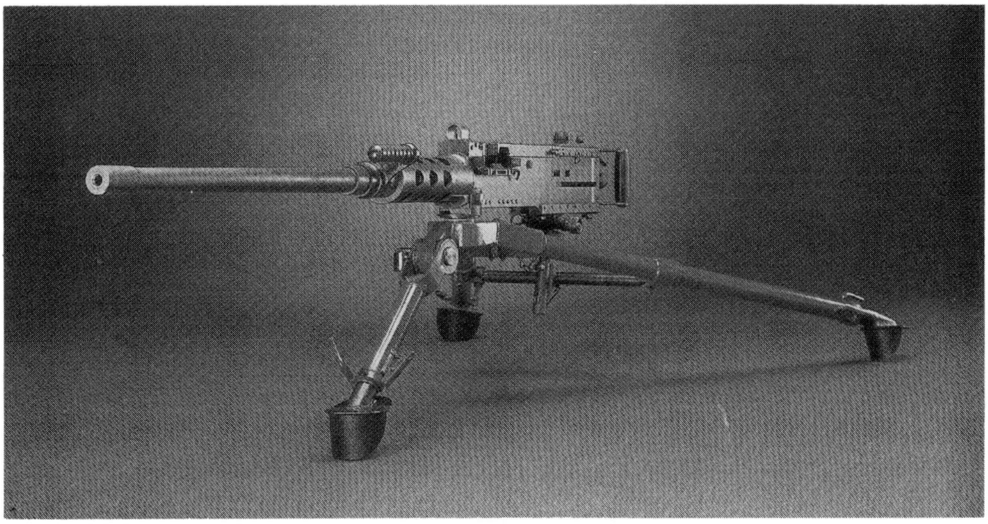

FIG. 2.8 0.5in Browning heavy machine gun

The method of breech locking for the 0.5in. Browning heavy machine gun is by protruding lug: it is shown in Figure 2.9. This protruding lug locking system requires very little space and so does not significantly increase the length of the receiver.

The advantages of the short-recoil operation are:

- Positive locking of the breech bolt.
- Lightweight breech bolt.
- Higher rates of fire than long-recoil operated weapons.

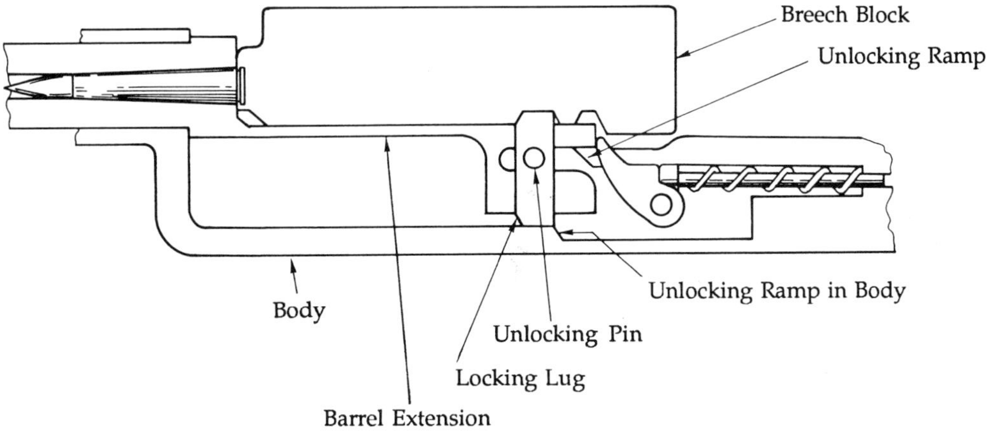

FIG. 2.9 Protruding lug breech locking

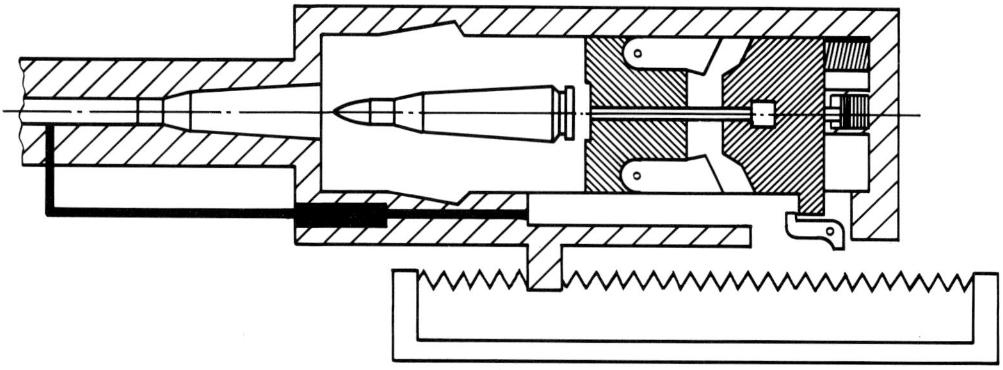

FIG. 2.10a The cannon is ready to fire, the bolt is retained by the sear and the return spring is compressed. A cartridge is ready to be fed into the chamber

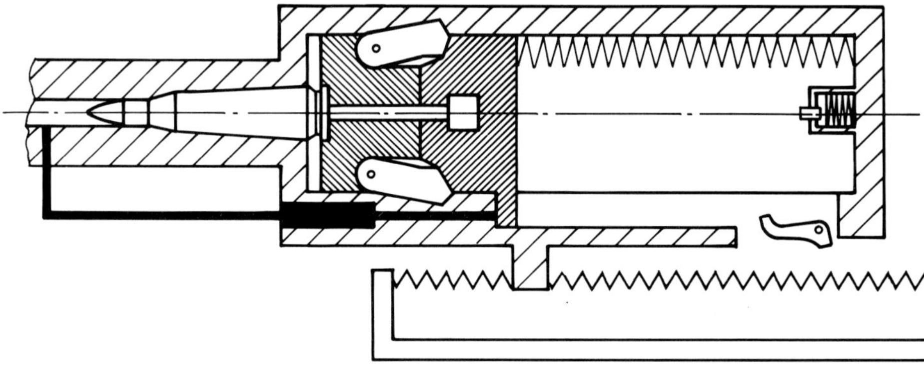

FIG. 2.10b The trigger is released and the bolt feeds a cartridge into the chamber. The breech bolt locks to the barrel, the firing pin strikes the primer and the cartridge is fired

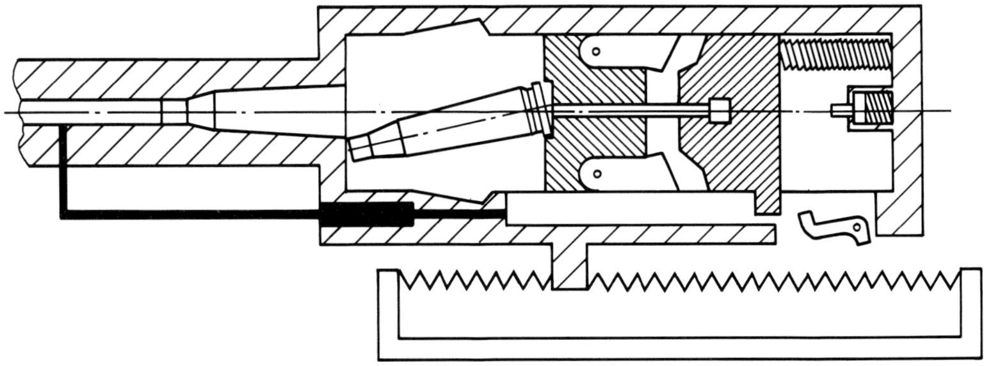

FIG. 2.10c After the projectile passes the gas port, the gas piston unlocks the breech block from the barrel. Residual gas pressure accelerates the breech block after the bolt has unlocked. The empty cartridge case is extracted and ejected

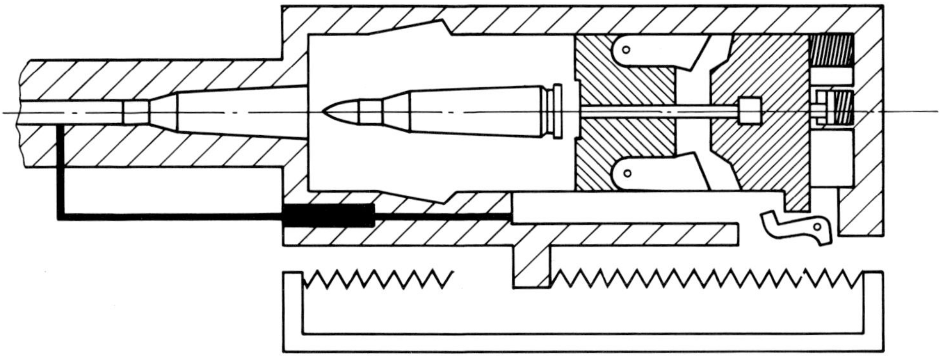

FIG. 2.10d A fresh cartridge is chambered when the bolt is pushed forward by the breech block return spring

The disadvantages of the short-recoil operation are:

- Long receiver.
- Complex design.
- Large and expensive barrel bearings are required.

Gas-Operated Cannon

The sequence of operations for gas-operated cannon is shown in Figures 2.10a to 2.10d.

Gas-operated cannon use the propellant gases to drive a piston which operates the mechanism to cycle the weapon. The propellant gases are bled from the barrel through ports drilled through the barrel wall and fed through channels into a gas cylinder in which is mounted a gas piston. The rearward moving piston unlocks the breech bolt from the barrel and pushes the breech block to the rear of the weapon. At the moment of unlocking, the gas pressure in the chamber is still fairly high so that the rearward movement of the breech block is assisted by blow-back action. The unlocking process can only be initiated after the projectile has passed the gas port, so that this type of system has automatic hang-fire safety built into it. This type of operation is capable of high rates of fire and so is widely used for anti-aircraft weapons.

Figure 2.11 shows the pivoting lock mechanism which is commonly used in long-established designs of medium-calibre, gas-operated cannon. The sliding block assembly is pushed away by the gas piston via the piston rod until the pivoting locks are free to swing in. Unlocking is completed when the sliding block assembly strikes the bolt head and sets the whole bolt in motion. The bolt is then accelerated by the residual gas pressure acting on the inner base area of the cartridge case and hence to the breech block. The rate of fire is thus determined to a large extent by the speed of the unlocking process, which, in turn, is determined by the pressure at the gas port and in the gas cylinder and by the distance from the breech face to the gas port. This type of locking mechanism is not often used on newly-developed weapons because it is necessary to use large, high-strength forgings for the body of the weapon to resist the high forces

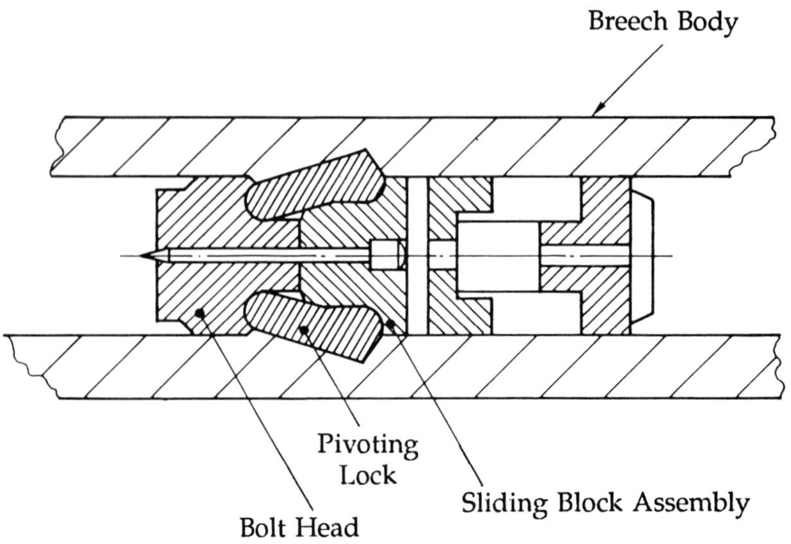

FIG. 2.11 Locking mechanism for pivoting lock bolt

exerted by the bolt head. Multi-lugged, rotating, locking bolts are almost always used on modern designs (see the sections on Chain guns and Gatling guns) because only the lugs and barrel extension in them are required to resist the high forces imposed upon the breech bolt by the high-pressure propellant gases.

As the barrel and gas parts heat up there is less heat loss from the propellant gases and so the pressures are higher, so rates of fire on gas-operated cannon increase with barrel temperature. Almost all gas-operated weapons rely on blow-back assistance to cycle the weapon and this is a major reason why they achieve the high rates of fire associated with this method of operation. However, when the breech is opened, high concentrations of toxic gases may be present. Because of this, gas-operated weapons are not often fitted into the turrets of light armoured vehicles.

The advantages of gas-operated cannon are:

- High rate of fire.
- Barrel and breech block are locked at the time of firing.
- Will operate with a mix of ammunition types producing varying gas pressures.

The disadvantages of gas-operated cannon are:

- Multi-component breech block.
- High thermal shock loads on gas drive mechanism.
- Rate of fire varies with weapon temperature.
- Fouling of piston assembly and bolt slides by propellant.

Revolver Cannon

Revolver cannon are a specific type of gas-operated cannon that were first developed during the latter stages of World War II. They were developed for use in aircraft because they can achieve very high rates of fire. The basic design uses a standing fixed breech; thus they are both light and compact, which are additional advantages for aircraft use. High rates of fire can be achieved because revolver cannon significantly reduce the time required to feed ammunition into the chamber. This is achieved by feeding ammunition in stages over a number of firing cycles. Figure 2.12 shows a typical arrangement for a revolver cannon.

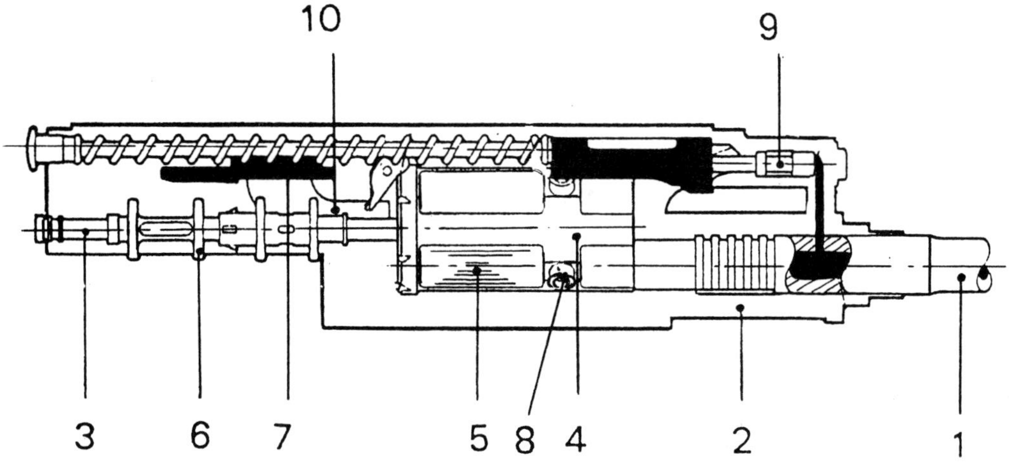

FIG. 2.12 Typical revolver cannon

The revolver cannon consists of a barrel (1) and a receiver assembly (2), which contains the complete feed mechanism. In the receiver assembly is a shaft (3) on which are mounted a cylinder (4), with four or more chambers (5), and the sprockets for conveying the rounds (6). In the top of the receiver is a control slide (7) with a cam slot in which the control slide moves backwards and forwards. The control rollers (8) are mounted around the periphery of the cylinder and engage in this slot. The control slide is driven by the propellant gases via a piston (9) which is fed from ports in the barrel. The reciprocating movement of the control slide acting in the cam slot turns the cylinder one-fifth of a turn for a weapon with five chambers per cylinder. The ammunition belt is drawn forward during the control slide's rearward motion, and, as the slide moves forward again, the two rammers (10) chamber the cartridge. The receiver is mounted in a cradle and recoils with each round fired.

Figure 2.13 shows the time/displacement diagram for a typical revolver cannon. Points A to G in the cycle indicate:

A. The cartridge is fired and the control slide is pushed to the rear.
B. The cylinder starts to rotate and ammunition feed begins.

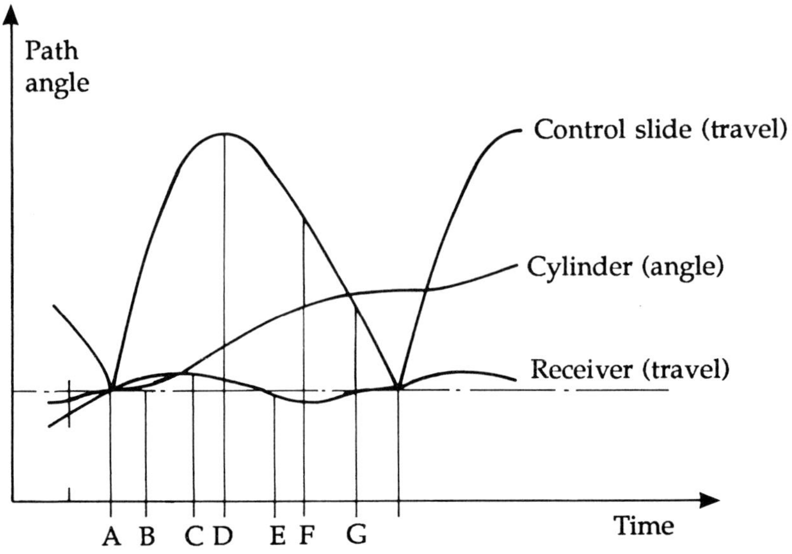

FIG. 2.13 Time displacement diagram for a typical revolver cannon

C. The receiver has reached the maximum recoil position.
D. The control slide has reached the rearmost point of travel.
E. The receiver has reached the point of rest and overshoots slightly.
F. Commencement of feed of a fresh cartridge into the chamber.
G. The receiver is at rest and feeding of the cartridge is completed. The remainder of the control slide's forward motion is used to eject the spent case.

High barrel wear is a particular problem with revolver cannon, especially with weapons firing high velocity ammunition. Typical rates of fire are in excess of 1,400 rounds per minute with the latest designs achieving over 2,000 rounds per minute. All of the rounds pass through the one barrel so that barrel heating, and thus wear, is high. This design of cannon also suffers from an additional form of wear known as 'balloting' which is caused by a high circumferential component of velocity imparted to the cartridge by the rotating cylinder. As the round is fired this causes the projectile to bounce off one side of the barrel and on to the other. This is repeated all the way down the barrel causing additional wear where it occurs. These high circumferential forces can cause the cartridge case crimping to loosen prior to chambering so ammunition used in revolver cannon need this to be strengthened. The reason that revolver cannon can achieve such high rates of fire is that ammunition is fed into the rotating cylinder in three stages. Thus to load a cartridge fully into the chamber ready to be fired requires three complete cycles of the weapon. External power must be used for this. Compressed air is often used and the air bottle typically has sufficient capacity for three complete loading cycles.

The advantages of revolver cannon are:

- High rate of fire.
- The fixed breech makes for a very compact design.
- Design is light in weight for the specific muzzle power.

The disadvantages of revolver cannon are:

- High barrel wear.
- Large eccentric inertia forces occur so that the barrel requires additional support.
- External power source is required for loading.
- Ammunition loaded in the hot cylinder between firings is susceptible to cook-off.

Conclusions

Self-powered weapons tend to be complex and expensive to manufacture but have reached a high level of development and reliability. Each different mode of operation has specific advantages which makes it particularly suitable for specific applications. Such weapons do not risk power failure if their parent vehicle is damaged and may still, for this reason, be used in the land-based role for some time yet.

3.

Cannon Using External Power

Advantages and Disadvantages

Many modern medium-calibre cannon designs are based upon operating systems which use external power rather than part of the energy developed by the ammunition. The resulting cannon are light and simple and have proved to be highly reliable, both mechanically and operationally. Increased reliability is brought about by a number of factors, one of the most important being the simpler mechanical design and the reduction in the number of working parts that this confers.

When an external power source is used it is not necessary to have an energy-storage device, such as springs or accumulators, with all of the high forces and complexity that they involve. With an externally-powered cannon it is also a simple matter to cycle dummy rounds through it on a continuous basis without firing expensive ammunition on an expensive range: so most aspects of the operating cycle can be tested in the workshop without firing a shot.

Externally-powered cannon can, if it is wished, continue to cycle even when a round does not function. The faulty round is extracted as if it had been fired and is ejected as if it were an empty case.

There are several disadvantages to using external power, the most obvious being that an external power source must be available for use. Such weapons are therefore usually confined to being mounted on vehicles and use the power source of the vehicle. A potential disadvantage is that if a hang fire occurs the cartridge may detonate with the breech in the unlocked state. Some designs have a built-in sensor to detect either the recoil or the pressure pulse in the barrel and prevent the breech unlocking if any delay in either occurs. These devices, however, often reduce the rate of fire of the weapon. Many users rely on increased standards of quality control to reduce the possible incidence to an acceptable level. Externally-powered cannon can also be double loaded, which is particularly hazardous with HE rounds.

Gatling Cannon

Gatling cannon is the name used for multi-barrelled cannon in which the barrel group rotates. The name originates from Dr R J Gatling who invented this concept in 1862. He applied the operating system to small-calibre weapons which were

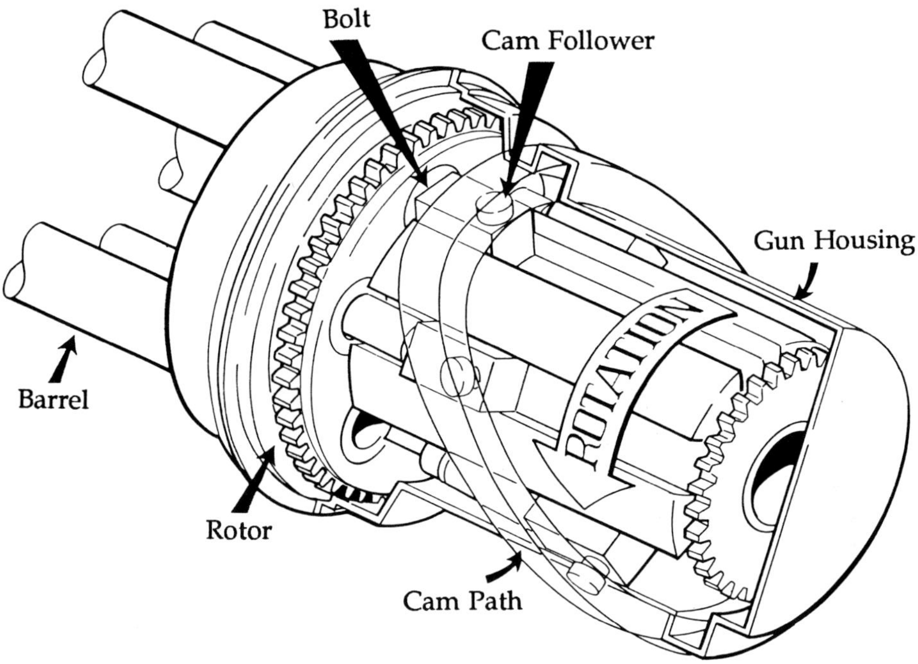

FIG. 3.1 Powered Gatling cannon rotor assembly

the forerunners of the machine gun. These weapons were also externally powered, being hand cranked by the firer. Modern Gatling cannon were developed in the West by the General Electric Company, now known as Martin Marietta. The main components of weapons operating on this principle are the rotor, receiver and the drive motor. The rotor consists of the barrel group and their locking breech bolts. There may be from three to seven barrels which are attached at the breech and the muzzles. The receiver is fixed and has a cam slot which controls the movement of each of the locking bolts. The drive motor provides the external power and may be electric, pneumatic or hydraulic.

The rotor, to which the barrels are fixed, rotates and each barrel is fired in turn at the same position with each complete revolution of the rotor. Each locking bolt is controlled by the cam slot which converts circular motion into reciprocating motion, the cam slot being fixed relative to the receiver. The forward motion of the bolt moving in the cam slot picks up a round and chambers it, the bolt locks into the barrel extension, the round is fired and the bolt unlocks and extracts the empty case from the chamber by the bolt moving to the rear, guided by the cam slot. Figure 3.1 shows the rotor assembly with the cam slot controlling the bolt movement for a typical Gatling cannon.

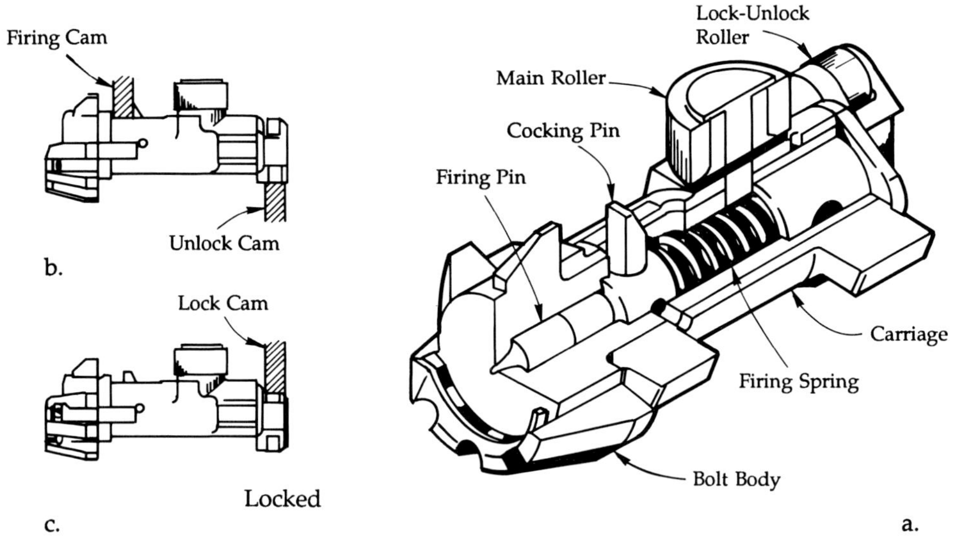

FIG. 3.2 Gatling cannon rotating bolt

Gatling Cannon Bolt Design

The locking bolts on the first Gatling cannon had protruding lugs, but the later designs have rotating, multi-lugged breech bolts as shown in Figure 3.2. Each bolt is driven along its rotor track by a roller travelling in the main cam slot which is machined into the body of the receiver.

The bolt assembly consists of the bolt body, a carriage, a main cam follower roller, lock – unlock rollers, a firing pin, a cocking pin and a firing spring. The carriage rides on tracks in the main rotor. The main roller engages in the cam slot in the receiver and drives the carriage from the rear dwell to the forward dwell position and back again for every complete revolution of the barrel group. A cam follower with the lock – unlock rollers is attached to the rear end of the bolt body. In the forward dwell position these rollers pass through a lock – unlock cam to rotate the bolt head to lock it into the barrel extension before the round is fired, and then rotates the bolt to the unlocked position after the round has been fired.

The firing mechanism consists of a firing pin, cocking pin and firing spring contained within the breech bolt, and the firing cam which is attached to the main receiver. The stationary firing cam cocks and releases the firing pin as the bolt assembly is driven forward and rotated into the firing position. After firing, the bolt assembly goes through the extraction cycle where the fired case is removed from the chamber and returned to the feed system. It should be noted that when the firing pin is released it is entirely under the control of the firing spring and not locked into the operating cycle. If the spring or the firing pin stick there may be a late strike of the primer and the round may be fired slightly out of sequence: this is a mechanical hang fire. Once the bolt has started to unlock, the firing pin is prevented from reaching the primer. However, a late primer strike can be dangerous because there is always a delay between primer strike and propellant ignition so that the bolt could be unlocking whilst the chamber pressure is high or even at the commencement of propellant ignition. The period for this to occur in is very short so that although such an occurrence is possible it is not very likely.

The rotating barrel group has considerable momentum so that when firing ceases there is a finite length of time for the barrel group to slow down and stop. During this time live rounds will still be fed into the cannon. They will not be fired but will pass through the weapon on to the fired side of the weapon. A reversing technique is used to clear this live ammunition from the cannon after each burst. An electro/hydraulic drive system controls this clearing function and ensures that the first live round is positioned consistently in its correct position on the live side of the weapon. Rounds that were cycled through the weapon but not fired at trigger release are returned to the live side of the feed system. All of this takes less than one second before the weapon is ready to fire again.

Figure 3.3 shows the operating cycle for one barrel of a typical Gatling cannon during one complete revolution of the barrel group. It can be seen that there are no sudden peaks of movement as associated with self-powered cannon. These peaks are not necessary because of the constant power which is available throughout the operating cycle. There are therefore no high accelerations and high forces to stress the firing mechanism unduly. Barrel stresses and barrel wear are also low for a weapon with such a high rate of fire because the rate per barrel is low.

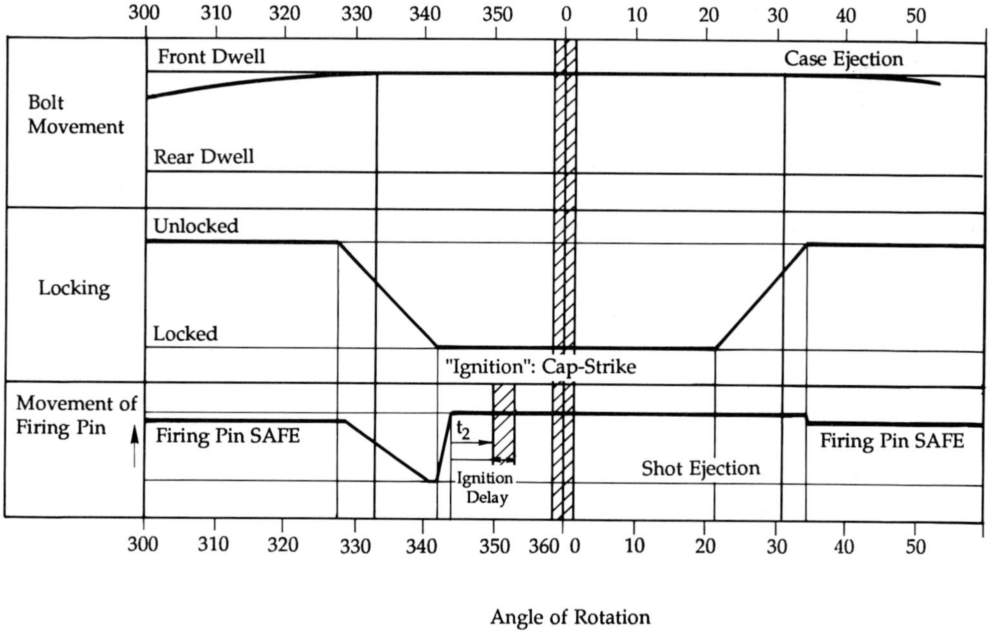

FIG. 3.3 Operating cycle for a typical Gatling cannon for one barrel

The dispersion of Gatling cannon fire is usually greater than that of single-barrelled weapons because of the mechanical problems of aligning all of the barrels to the same point of impact. In addition, it takes typically 0.5s to reach full rate of fire from the moment that the trigger is pressed. During this time the accuracy is reduced because the barrel group is still speeding up. Rotational velocity of the barrels for a typical 30mm cannon is in the region of 10m/s. Thus for the first round fired with the barrels stationary, or nearly so, the projectile will go where the barrels are pointing. At full speed the projectiles will have a velocity of 10m/s at right angles to the direction that the barrels are pointing. For a muzzle velocity of 1,000m/s this will result in approximately 10 milliradians deviation from the line of the barrels, which is constant when the barrels have reached full speed.

The actual cannon mechanism is very simple in the Gatling cannon: it is the feed mechanism that is required to satisfy their huge demand for ammunition that adds considerably to the weight and complexity of the complete system. Figure 3.4 shows the GU-8 Gatling cannon only, from which it can be seen that the actual cannon is remarkably compact with a weight of only 280kg. However, with the feed system and ammunition storage the total weight, including the 1,171 rounds of ammunition, is 1,723kg. The power required to drive the whole system is 77hp.

The advantages of the Gatling principle are:

- Capable of very high rates of fire.
- Low barrel heating and barrel wear.
- Very simple mechanical design for the cannon.
- Design is light in weight for the specific muzzle power.

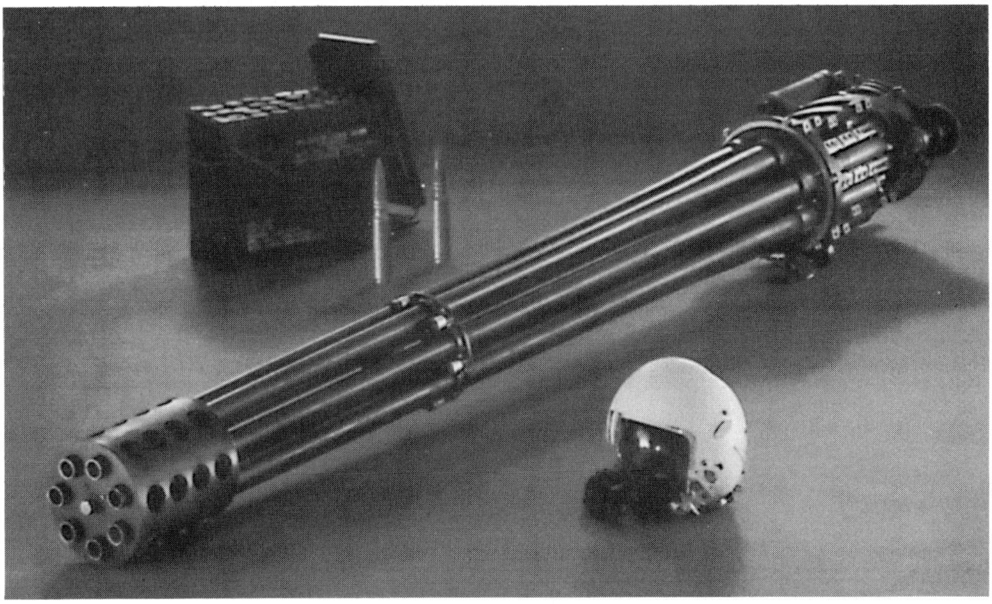

FIG. 3.4 GU-8 30mm calibre Gatling cannon (Courtesy: Martin Marietta)

The disadvantages of the Gatling principle are:

- Requires a large external power source.
- The feed mechanism is large and complex.
- Requires large quantities of ammunition to feed the high rates of fire.
- Relatively large dispersion.
- Shift in the mean point of impact during acceleration of the barrel group from rest.

Salvo-Firing Cannon

The earliest method of increasing the rate of fire of a weapon system was to raise the number of barrels and then fire each in turn, in groups or all together. These very early salvo-firing weapons were medium- and small-calibre muzzle loaders with glowing-match ignition, which can be traced back to the fourteenth century. This design was adopted because of the very long time required to reload muzzle-loading weapons. Salvo-firing weapons in different forms were produced into the 19th century, with as many as 80 barrels formed into one group and with the use of percussion ignition. Channels linked groups of barrels so that one percussion cap could fire separated clusters of barrels. The advent of the metallic cartridge case, which allowed the rapid reloading of a single barrel, superseded the salvo approach until the need arose for ultra-high rates of fire which were so high that it was physically impossible to load ammunition into a single barrel and remove the empty case in the very short time between shots. In addition, chamber pressure would also be very high as the case was being extracted so that the unsupported case would rupture. One answer was the multi-barrelled Gatling gun, discussed in the previous section. Another solution was a return to the salvo approach with multi-fixed barrels firing cased ammunition to allow rapid reloading between salvoes.

The only modern salvo-firing weapons are the 20mm and 30mm calibre Meroka which were developed in Spain for use against low-flying targets. They have 12 barrels, in two banks of six, arranged above one another. A separate cartridge belt feeds ammunition to each bank. The barrels are linked to one another at the front and combined into a common barrel block at the rear. The adjoining housing contains the breech body common to all barrels, the pawls for transporting the belt, and the locking devices.

The salvo, consisting of one round per barrel, is fired at a rapid rate in six double shots. Spacing the timing of the firing in this way reduces the recoil forces transmitted to the cannon mount. The firing pins are electrically operated. After the last shot, reloading begins automatically, using compressed air. The spent cases are drawn again into the belts by the recoiling breech block, the belts are moved on by six positions, the advancing breech block chambers the cartridges and is then locked. Automatic fire can then continue with the next salvo. The time required for reloading up to the moment when the gun is ready to fire is 0.2s for the 20mm calibre weapon and 0.25s for the 30mm calibre weapon. The chronological cycle of events is shown in Figure 3.5.

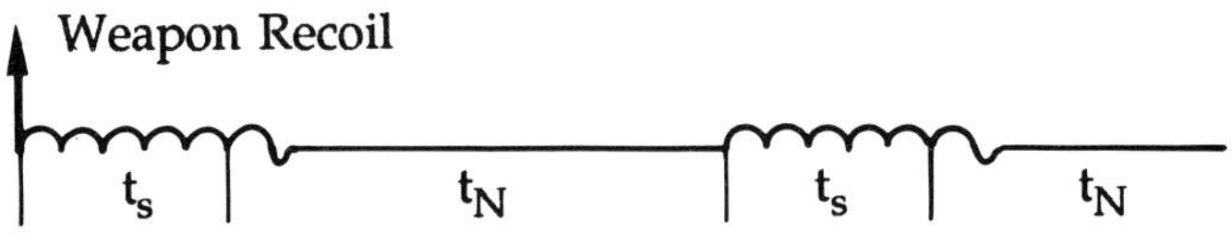

t_s = 0.06s intervals between salvos of 6 double shots

t_N = Reloading time

FIG. 3.5 Firing cycle for a salvo firing cannon

FIG. 3.6 Meroka salvo-firing cannon (Courtesy: Santa Barbara SA)

The Meroka salvo-firing cannon has been developed specifically for mounting on ships and for land use. Weight is therefore not such a critical factor as it is in Gatling cannon, which were originally developed for aircraft. Thus the heavy breech arrangement associated with salvo-firing weapons is acceptable. Dispersion can be adjusted by altering the spacing of the barrels at the muzzle. It is claimed that dispersion during a salvo is less than 2 milliradians. Because of the fixed barrel arrangement, there are no problems of initial inaccuracy as there are with rotating barrel guns.

Figure 3.6 shows the Meroka multi-barrelled naval air-defence weapon.

The advantages of the salvo-firing system are:

- Capable of very high rates of fire.
- Low barrel heating and barrel wear.
- Simple mechanical design.
- Low dispersion.

The disadvantages of the salvo-firing principle are:

- Heavy breech block.
- Requires an external power source to operate.

Chain Gun

Hughes Helicopters (now McDonnell Douglas) began developing chain guns in 1972 and have since built up a family of weapons ranging from 7.62mm to 30mm calibre. Weapon cycling and ammunition feed are by an electric motor. The rate of fire depends upon the choice of motor speed, which is governed by the supply voltage. The weapons are single barrelled and the rates of fire are in the same region as conventional cannon, up to 700 rounds per minute. The weapon is lightweight in design and mechanically simple. The cycle of movements is controlled by an endless chain, hence the name for this type of weapon.

Drive System

At the heart of the system are four sprockets which are coupled together by a duplex endless chain. An electric motor provides power to one of the sprockets to power the whole weapon. A sliding member projects from the chain and engages in a slot in the bolt carrier. Thus circular motion is converted to reciprocating motion and the bolt carrier is driven backwards and forwards by the chain. Figure 3.7 shows the basic arrangement used. The dwell time required for the period when the bolt is locked and the round is fired is obtained by the transverse movement of the sliding member. The dwell time required when the bolt is at the rear and the ammunition feed is activated is again obtained by the transverse movement of the sliding member, but this time moving in the opposite direction.

Locking of the breech block to the barrel is achieved using a multi-lugged rotating bolt. This considerably reduces the size, weight and strength

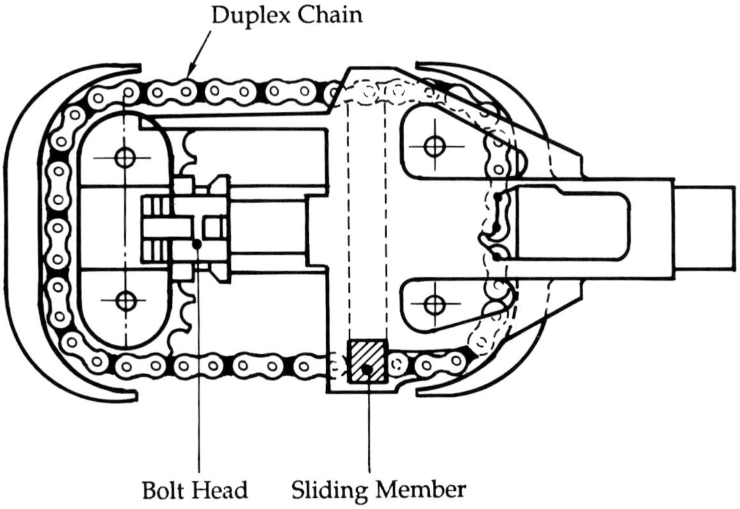

FIG. 3.7 Operating principle of the chain gun

requirements of the receiver because all of the loads from the propellant gases are applied only to the lugs of the locking bolt and a short barrel extension.

Like all externally-powered weapons, chain guns may be damaged by hang fires, whether caused by the ammunition or the gun. This problem is eliminated by an interlock which stops the weapon before the bolt unlocks if the weapon does not recoil. However, this safeguard limits the rate of fire of the weapon. The practical maximum rate of fire with a recoil sensor is about 400 rounds per minute. For higher rates of fire the hang-fire protection is ineffective.

Chain Gun Feed System

The main differences within this family of cannons are limited to only the calibre and the number of feed mechanisms. The dual-feed system of the 25mm calibre M242 fitted to the Bradley Infantry Fighting Vehicle uses two separate feed sprockets and a system of clutches so that either of the feed sprockets can be engaged to feed the selected cartridges. The feed select solenoid is also incorporated to position the feed sprockets between bursts so that the clutches are properly aligned for shifting. The feeder interfaces with the reciprocating bolt assembly through an intermittently-driven ammunition-transfer rotor. This rotor takes ammunition from either constant velocity feed sprocket and places it within the fixed extractors on the bolt face, ready for ramming. After firing and extraction, the transfer rotor sweeps the fired case from the bolt face. By modifications to the gun mechanism the case can be made to eject forward or sideways. When the cartridge first enters the feeder, it passes a link aligner on the link strippers that ensures correct alignment of the round. After passing the link aligners the rim of the round is fed into a T-slot to retain it in the correct axial position. When the cartridge reaches the ram position it is transferred to the T-slot in the bolt face.

A cross-sectional view of the two paths in the feeder is shown in Figures 3.8 and 3.9. Figure 3.8 shows the sequence of events for the upper feed as follows: the feed sprocket, which is driven directly by the main motor, transports the belted ammunition into the gun at a constant velocity through a pair of link strippers. As the cartridges are side-stripped from the links, they are guided out of the feed sprocket and placed successively into the presented cavity in the feed rotor. The transfer rotor remains stationary while the bolt rams, fires and extracts the previous round. The transfer rotor is then rotated by the index drive and the fired case is swept from the face of the bolt and a new round from the feed sprocket is positioned in its place.

Figure 3.9 shows the sequence of events for the lower feed, the operation being identical to the upper feed path. It should be noted that the stripped links are ejected in a different position for each feed path. The purpose of having two paths is to allow two different types of ammunition to be selected at the flick of a switch. With this type of arrangement it is not quite possible because, when switching from one feed to the other, there will always be one round already in the transfer rotor from the previous ammunition belt which must be fired before the newly-selected round can be fired.

Electrical power requirement for the 25mm calibre M242 chain gun is in the region of 8hp for rates of fire of 400 rounds per minute.

Figure 3.10 shows the 30mm M230 chain gun fitted to the Apache attack helicopter.

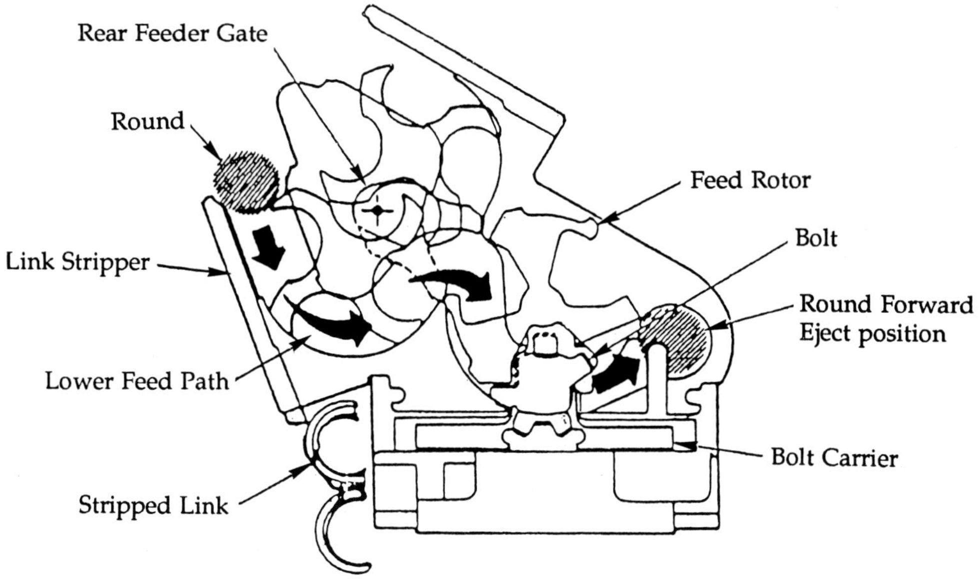

Fig. 3.8 Chain gun upper feed operation

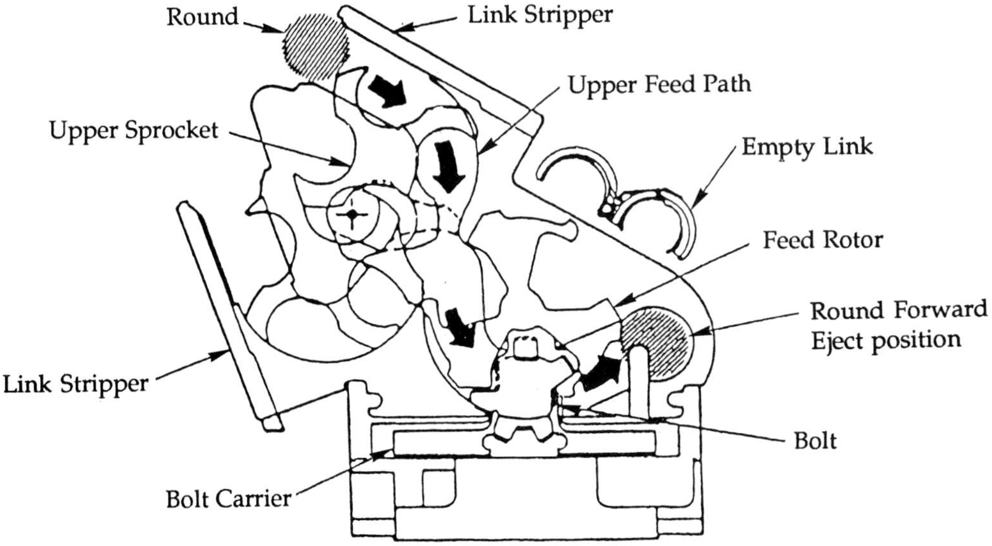

FIG. 3.9 Chain gun lower feed operation

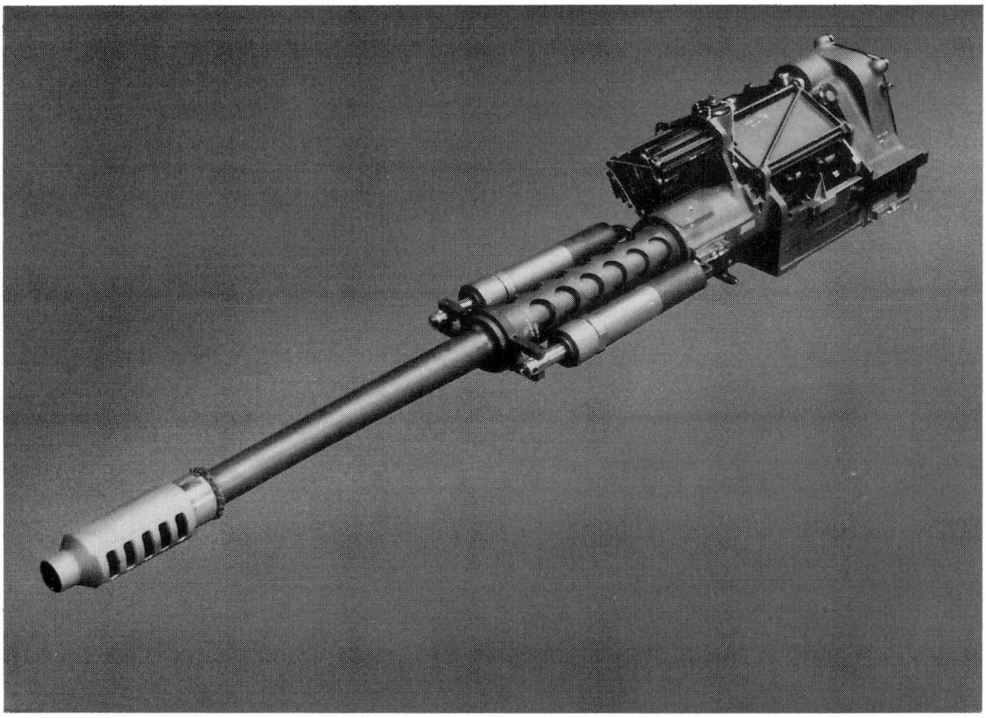

FIG. 3.10 30mm M230 chain gun (Courtesy: McDonnell Douglas)

The advantages of chain guns are:

- Low weight, short overall length and compact design.
- Variable rates of fire
- Low gas pressure upon cartridge extraction reduces fumes to the rear of the gun.
- Simple mechanical design results in high levels of reliability.

The disadvantages of chain guns are:

- Requires an external power source.
- Delay of two cartridges when changing ammunition natures.
- Ejection takes place in three directions with dual belt feed.
- High rates of fire not possible with hang-fire detection.

Rotating Element Cannon

A number of externally-powered cannon have been developed which convert rotating movement into reciprocating movement to drive the locking bolt. They do this by using a cam path cut into the rotating element. An extension to the base of the locking bolt runs in the cam path and controls the movement of the locking bolt. A typical example of this type of mechanism is the Rheinmetall Rh503 cannon. This was originally being developed for installation in the German Army's upgraded Marder Armoured Fighting Vehicle. This upgrade was cancelled but Rheinmetall have continued with the development of the cannon. Figure 3.11 shows a diagrammatic layout of the cannon.

Power is supplied by an electric motor; power consumption is 8 kW. This drives the rotating element through a reduction gear box. The cam path machined into the outer surface of the rotating element drives the breech bolt backwards and forwards and provides the pause in the forward locked position when the weapon is fired and that when the bolt is to the rear to allow the feeding of ammunition. As with most modern designs, the breech block is of the rotating type design and has two locking lugs. The weapon may be fired single shot or with an adjustable rate of fire of between 150 and 200 rounds per minute. To prevent

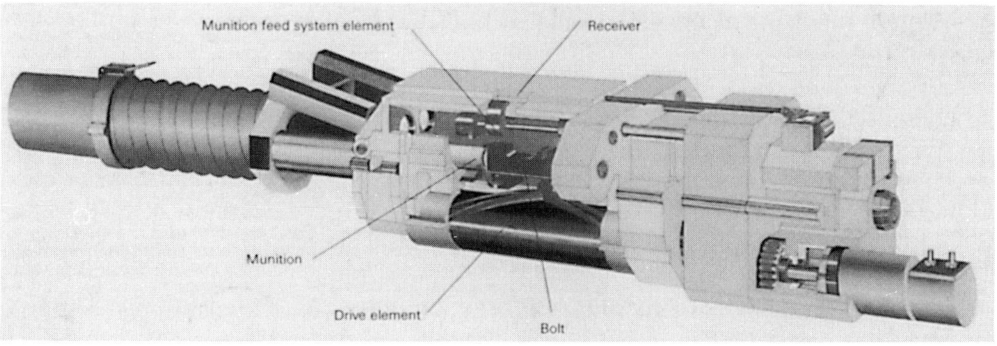

FIG. 3.11 Rheinmetall Rh503 gun mechanism (Courtesy: Rheinmetall GmbH)

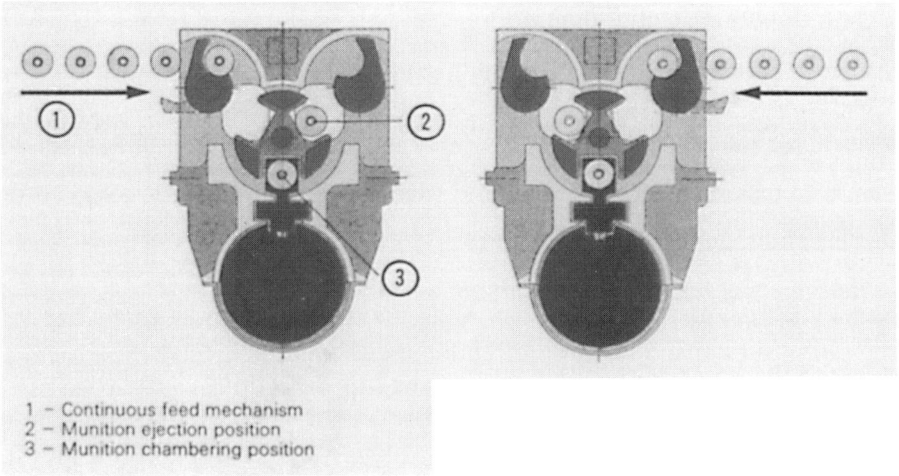

FIG. 3.12 Dual feed mechanism of the Rheinmetall Rh503 cannon (Courtesy: Rheinmetall GmbH)

breech explosions caused by hang fires, an interlock is fitted to prevent the bolt unlocking until after the weapon has fired. This interlock is gas operated by use of a gas port drilled in the barrel. Sensing the gas pressure allows a higher rate of fire to be achieved than when sensing the recoil movement. The weapon features dual ammunition feed. This is achieved by feeding ammunition from either the left-hand side or the right-hand side, as shown in Figure 3.12. The direction of rotation of the electric motor is reversed when a different type of ammunition is required to be fired. The advantage of this type of design is that when changing from one feed direction to the other the new ammunition is available immediately. The disadvantage of this method is that two ejection ports are required.

Advantages of the rotating element design are:

- Low weight, short overall length and compact design.
- Variable rates of fire.
- Low gas pressure upon cartridge case extraction reduces fumes to the rear of the gun.
- Simple mechanical design results in high levels of reliability.

Disadvantages of the rotating element design are:

- Requires an external power source.
- Ejection takes place in three directions with dual belt feed.
- High rates of fire not possible with hang-fire protection.

Conclusions

Most modern designs of medium-calibre cannon use an external power source to cycle the weapon. This results in a simpler and more compact design which usually proves to be highly reliable. The one obvious major disadvantage of these types of weapon is that they require a power source in order to cycle them. However, because most of these weapons are vehicle mounted they are able to use the vehicle's power source. A failure of the external power source may result in the failure of the weapon to cycle, although IFV cannon can usually be hand cranked at about two rounds per minute.

4.

Barrel Design

The Need for Rifling

All modern medium-calibre barrels are rifled to impart spin to the projectile, so giving it gyroscopic stability. The trend is towards fin-stabilised projectiles for the primary round of IFV cannon in order to achieve the required armour penetration. However, the primary round is so called in terms of absolute performance. The round which is used as frequently, or more so, is the full-calibre HE round. Because of this, the rifled barrel is likely to remain the norm for future medium-calibre weapons rather than a change to a smooth-bored barrel as has happened with tank guns. Cannon systems tend to increase muzzle velocities to improve armour penetration. The resultant reduced flight time of the projectile increases the hit chance at the target. Thus the barrel length is usually maximised to maximise muzzle velocity.

Rifling Form

A rectangular rifling profile is normally used, as shown in Figure 4.1. Whilst other rifling profiles have been used, experience has shown that they confer insufficient advantage to warrant the additional machining costs that are usually incurred. Rifling groove depth, the number of grooves, groove width and land width all have a considerable influence upon barrel life and have been established empirically after many years of experience with different rifling forms. Typically, the rifling depth, t, is 2 per cent of the calibre of the weapon. Thus for a 30mm calibre cannon the rifling depth would typically be 0.6mm. Land width is a compromise between the need for lands to be as wide as possible to minimise barrel wear but as narrow as possible to minimise the engraving force applied to the driving band. It has been found that the best compromise is to make the grooves twice the width of the lands.

The number of grooves is usually approximately 60 per cent of the calibre of the weapon in millimetres. Thus a 30mm weapon would usually have 18 rifling grooves. It is also usual to have an even number of grooves. It is again usual for the rifling twist to be right-handed (clockwise) viewed from the breech towards the muzzle.

The angle of twist is decided upon by the requirement to stabilise the projectile gyroscopically. A 30mm calibre weapon will typically have a twist rate at the muzzle of one turn in 800mm, which for a muzzle velocity of 1000m/s will give

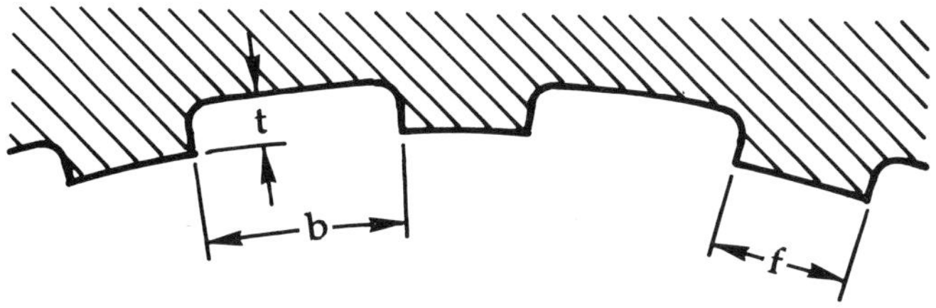

t = Depth of groove
b = Width of groove
f = Width of land

FIG. 4.1 Diagram of groove profile

a spin rate of 75,000 revolutions per minute. Medium-calibre cannon are the only weapons that regularly use non-uniform twist rates for their barrels. With constant (uniform) twist the angle of the rifling is constant from the chamber to the muzzle, which is usual for most small arms, artillery and tank guns. The majority of medium-calibre cannon use gain twist, sometimes known as progressive, rifling in which the value of twist starts at a shallow value at the breech end and increases to a maximum at the muzzle end. Constant twist rifling has the advantage of being relatively simple to machine but produces high peak loadings on the lands and driving bands as the round engages in them. Gain twist rifling reduces the peak loading but is relatively more complex to machine.

With all forms of rifling the angular velocity, angular acceleration and angular accelerating force acting on the projectile are all linked to the linear velocity, linear acceleration and linear acceleration force: they are linked by the twist rate. If the twist rate is constant then the angular accelerating forces will be directly proportional to the linear accelerating forces. Varying the twist rate will modify the relationship between angular and linear velocity and allow the high peak accelerating force acting on the driving bands to be reduced. However, an additional load is exerted on the driving bands by the wedging action of the increase in twist rate, so that the advantage in using progressive rifling is not as high as would first appear.

The two occasions on which progressive rifling confers a real advantage is in revolver cannon and when the barrel is badly worn. In some designs of revolver cannon the projectile travels a relatively long distance in the smooth-bored cylinder before engaging the rifling in the barrel. Thus it has a high linear velocity when it starts to engrave the driving band. If the twist rate is constant the initial angular acceleration force is very high and results in shearing off the driving bands. Gain twist rifling with zero initial twist rate allows the angular acceleration of the projectile to be much more gentle so that the driving band is not damaged. The same concept applies to a badly-worn barrel where the initial part of the lands has been worn away.

Angular Acceleration

The largest force acting on the driving band, and which may cause the driving band to be sheared from the body of the projectile, is the force which gives angular acceleration to the projectile via the rifling. Figure 4.2 shows the forces acting on the body of the projectile as it is accelerated down the barrel by the propellant gases.

The acceleration force on the driving bands which gives the projectile its angular velocity is directly proportional to the chamber pressure. Thus the plot of the acceleration force on the driving band with respect to the distance travelled down the barrel will be similar in shape to the pressure/distance curve, i.e., a high initial peak followed by a gradual fall towards the muzzle. Different rifling twist rates are designed to eliminate this initial high peak. The mathematical proof is given in the Annex to this chapter.

Progressive twist rifling reduces the peak force on the driving bands, but, depending upon the initial twist rate and the rate of increase, the force will usually be higher towards the muzzle than with constant twist rifling. Thus, as

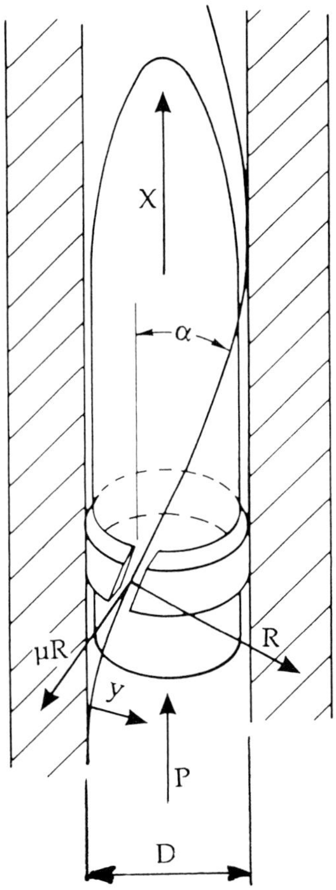

P	=	linear accelerating force due to propellant gas
α	=	angle of twist
X	=	direction of shot travel
y	=	circumferential movement of the driving band
R	=	angular accelerating force on the driving band
μ	=	coefficient of friction
D	=	driving band diameter

FIG. 4.2 Forces acting on the body of the projectile

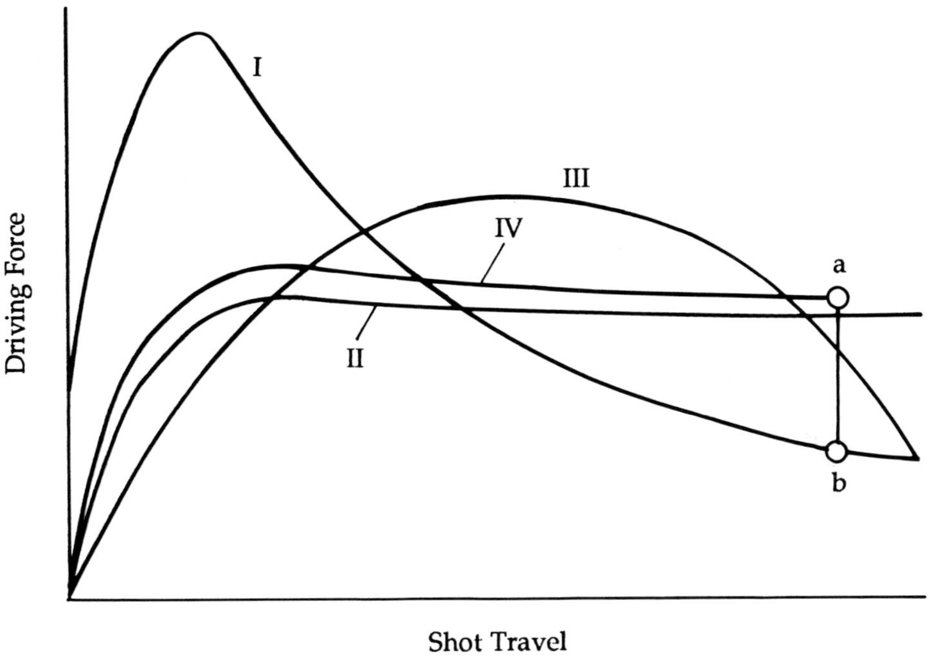

I = constant twist — $y = x \cdot \tan \alpha$
II = parabolic twist — $y = c_1 \cdot x + c_2 \cdot x^2$
III = cubic–parabolic twist — $y = c_1 \cdot x + c_2 \cdot x^2 + c_3 \cdot x^3$
IV = parabolic twist with constant twist at the muzzle point a.

where:
c_1, c_2, and c_3 are constants.
x is the distance from the breech measured along the axis of the barrel.

FIG. 4.3 Driving band force plotted against shot travel for different rifling twist rates

the projectile leaves the muzzle there will be a high circumferential force being applied to it, which is detrimental to projectile accuracy. To reduce this possible effect the final section of the rifling is often made with a constant twist rate. If the propellant gas pressure at the muzzle is low then the driving band force will also be low for constant twist rifling. Figure 4.3 shows the force on the driving band with respect to the position down the barrel for different rifling twist rates.

Barrel Manufacture

Cannon barrels start as billets which are forged to near nett shape, usually on a rotary forge. The forging starts as a solid shaft which is rough machined on the outside before being deep-hole drilled. The bore is then reamed and honed to its final size which is the nominal calibre of the weapon. Machining of the rifling grooves is usually achieved by broaching. This is carried out in one pass using up to 60 cutting heads. As the cutting heads are pulled through the barrel they are turned to achieve the required twist rate. For constant twist rifling the heads are usually fixed relative to one another on a central shaft. For gain twist rifling only the first cutting head is fixed and the subsequent ones are designed to follow the groove cut by the leading head.

Small-calibre rifled barrels are usually cold-hammer forged. The process is used commercially for producing barrels up to 20mm calibre. Barrels up to 30mm calibre have been produced by this process but it is unlikely that such large calibres will be produced commercially. A disadvantage with cold-hammer forging is that only rifling of constant twist rate can be produced.

A recent development in the forming of rifling grooves is the use of electro-chemical machining. The process consists of removing material by electrolytic action. The barrel is deep-hole drilled in the normal manner and a plug through which the electrolyte is passed is placed in the bore. This plug is a brass rod with the reverse pattern of the rifling grooves formed in the outer surface. Plastic inserts provide insulation and restrict the area to where the material is to be removed. The electrolyte is forced under high pressure along the rifling grooves machined in the plug and metal is removed from the barrel blank where the electrolyte comes into contact with the surface of the barrel. Rifling depth is set by the process time and the current density. Using a plug of short length the rifling may be machined with any twist rate by pulling the plug through the bore and controlling the rotation by computer. By using different electrolytes almost any material may be machined providing it is electrically conducting. It is usual to machine one pair of rifling grooves at a time and to index the tooling through the appropriate angle to machine the other groove pairs.

Barrel Materials

Most cannon barrels are made from a low alloy steel, typically containing 3 per cent nickel, 1 per cent chromium, 0.7 per cent molybdenum, 0.2 per cent vanadium, and about 0.3 per cent carbon. It is usual for the steel to be double melted and electro-slag refined so that it is a very 'clean' steel with good fatigue properties. The usual final heat treatment is a normalising/grain refining one to ensure constant strength and good impact properties. Barrel bore hardness is usually quite low, typically being in the range 32 to 36 Rockwell. This material has proved to be the best all-round compromise of high strength, high toughness, reasonable hardness, reasonable cost and easy machining. Resistance to wear is also good. Some manufacturers nitride the bore of the barrel in an attempt to reduce the problem of barrel wear. Nitriding is a process in which a steel component is heated to between 500 and 550°C in a nitrogen atmosphere to allow nitrides to form on the surface of the steel with certain specially added elements. The resulting surface is very hard and is unaffected by subsequent heat treatment. The usual barrel steels do not have good nitriding properties so special nitriding steels are used. A typical composition is

3 per cent chromium, 1 per cent molybdenum, 0.2 per cent vanadium, less than 0.3 per cent nickel and less than 0.35 per cent carbon. The usual surface hardness after nitriding is 64 Rockwell. These steels are more difficult to machine than the normally used barrel steels. There is little evidence to suggest that there is any significant benefit from the use of nitriding, although the surface of the bore is very hard.

Hard chromium plating is used extensively to protect both small-calibre and large-calibre barrels to reduce barrel wear significantly. However, when used to protect the bore of medium-calibre cannon, chromium plating is less effective in reducing barrel wear. Consequently few cannon barrels have their bores protected by hard chromium plating.

Chamber Pressures

Modern medium-calibre cannon have high muzzle velocities. To achieve this requires high chamber pressures, which are typically in the region of 400 MPa or higher. Figure 4.4 shows the pressure/time curve for the 30mm calibre RARDEN cannon. It can be seen that peak pressure is reached over a period of only 2ms. Pressure then falls until 10ms after shot start, the point when the projectile leaves the barrel, the pressure is less than 20 MPa. The propellant is all burnt after the projectile has passed approximately one-third of the way down the barrel. At the muzzle the pressure is low. Barrel length for the 30mm calibre RARDEN is 3m. To make the barrel any longer than this would be of little benefit because the low propellant gas pressure at 3m would give little increase in muzzle velocity if the barrel were made longer.

Barrel Heating

Whenever a cartridge is fired in a gun there is a release of chemical energy in the form of heat, which is, in turn, converted into mechanical energy as the projectile is propelled up the barrel. Up to 30 per cent of the heat from

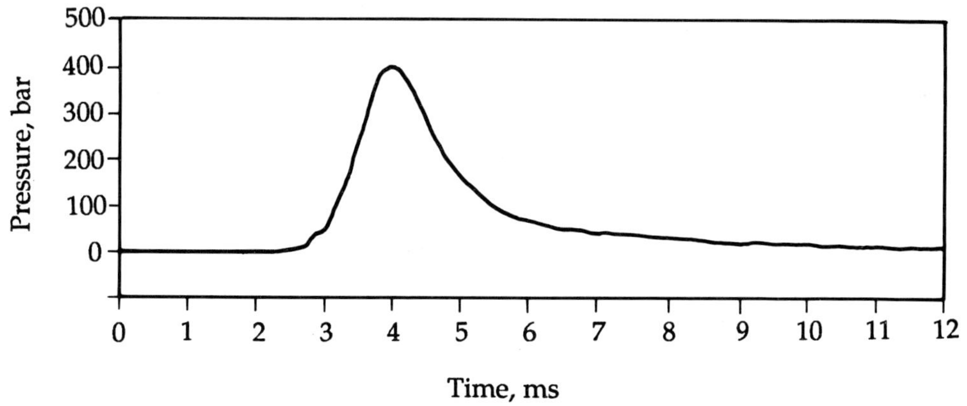

FIG. 4.4 Pressure/time curve for the 30mm calibre RARDEN cannon

the propellant is transferred to the gun barrel, which results in an increase in the temperature of the barrel. This increase is dependent upon the mass of the barrel: the thicker the barrel then the lower will be the temperature increase per round fired.

Typically the temperature rise will be between 2 and 5°C per shot fired. Barrel mass is usually greater for high rate of fire weapons, such as anti-aircraft weapons, to reduce barrel temperatures. Aircraft barrels, however, need to be light in weight. Because of this, and because they fire a large number of rounds in a short time, barrel heating and barrel wear is particularly acute for such weapons.

Cooling by radiation and natural convection is small and so has very little effect upon barrel temperature during firing. Consequently barrel temperature is largely dependent upon the number of rounds fired rather than the rate of fire. Multi-barrelled guns therefore have fewer problems with barrel heating and barrel wear than single-barrelled guns firing the same number of rounds. To overcome this problem of barrel heating, water cooling may be used as in the Russian ZSU-23 anti-aircraft gun or the naval version of the Bofors L70 anti-aircraft gun. Whilst this is a highly effective way to control barrel temperature, it adds considerably to the weight and complexity of the cannon system.

Barrel Wear

Barrel wear is a significant problem in guns which are either sustained fire weapons or have high muzzle velocities. Firing large quantities of ammunition in a short time results in the barrel being heated to very high temperatures which may exceed 400°C on the outside and significantly higher on the inside of the barrel. These high temperatures degrade the wear-resisting properties of the barrel material. Additionally, chemical attack by the burning propellant on the surface of the barrel bore significantly degrades the wear-resisting performance still further.

High muzzle velocities can only be achieved by using high energy propellants burning at high pressure. This combination leads to high propellant gas temperatures. Each time the weapon is fired it receives a high temperature pulse which lasts for a few milliseconds but which raises the barrel bore surface temperature to values as high as 1,000°C. Figure 4.5 shows the measured surface bore temperature for the 30mm calibre RARDEN cannon when firing the APDS round which is typical for a modern, high performance cannon. These high bore temperatures have a severely detrimental effect upon the ability of the barrel to resist wear.

Modern, high performance, medium-calibre cannon are usually sustained fire and have high muzzle velocities, a combination which leads to severe barrel wear problems. These problems are such that on some systems the barrel life is measured in hundreds of rounds. Such a short life is not only expensive in terms of barrel costs but also in terms of logistic support in the supply and replacement of barrels. So that operational performance is not compromised, barrels are usually replaced prematurely. This is especially the case for aircraft cannon barrels where it is imperative that they last for a full sortie.

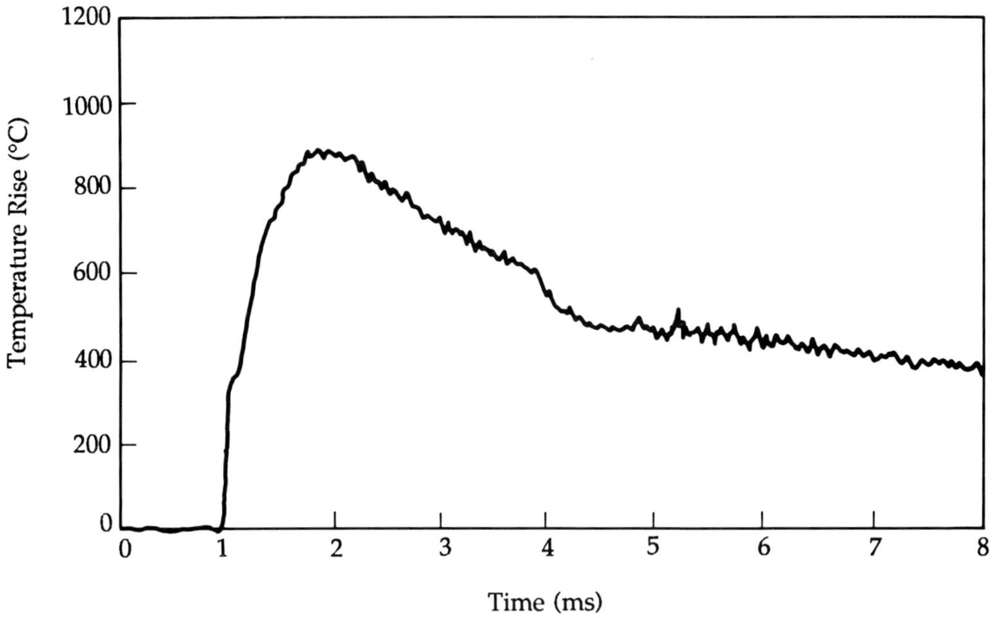

FIG. 4.5 Typical temperature/time curve at the surface of a 30mm calibre cannon

Barrel Condemnation Criteria

Barrel wear is greatest at the breech end and diminishes towards the muzzle. Wear is often referred to a bore measurement at 25mm from the commencement of rifling; it is at that point that wear is usually the greatest. However, the importance of the bore diameter is that it is an indication of the likely performance of the barrel. The effects of barrel wear are as follows:

- Reduction in the muzzle velocity.
- Bursting of the barrel.
- Loss of projectile stability.
- Increase in projectile dispersion.

Reduction in muzzle velocity is caused by gas leakage past the projectile, which is a result of an enlargement in the barrel diameter due to wear. Forward obturation of the propellant gases is provided by the driving bands, which are the same diameter as the rifling grooves. As the grooves wear, the driving band cannot fully seal the bore. There is no 'set up' of cannon ammunition as in small-calibre weapons, i.e. an expansion of the projectile caused by the high chamber pressure when firing. Thus small amounts of barrel wear may have a significant effect upon barrel sealing. The limit on the reduction in muzzle velocity is usually set at 6 per cent of that for a new barrel before a worn one is considered unserviceable. Reductions in muzzle velocities greater than this will result in projectile trajectories being affected sufficiently to reduce the hit chance significantly. It is rare for such a reduction in muzzle velocity to occur in practice because some other form of failure usually occurs before then.

Bursting of the barrel may be caused by a number of reasons not connected with barrel wear; these reasons will not be considered here. A possible cause of bursting due to large number of rounds being fired through it is fatigue. This is a mechanism whereby the continued stressing of the barrel, as it is fired, initiates surface cracks which grow with subsequent firings. A point is reached where the barrel is sufficiently weakened for it to burst. Such a failure mechanism is associated with thin-walled barrels and high chamber pressures. For many cannon barrels some other form of wear occurs before that of fatigue. Where fatigue is a problem it is usual to limit the number of rounds fired to a safe value to ensure that the barrel does not reach a dangerous state. For such barrels it is important to record all firings through it accurately, so it is usual to have a firing log book.

Loss of projectile stability is caused by imparting insufficient spin to it. When the barrel wears it is the rifling lands which are affected the most. A point is reached where the lands are worn away completely near the commencement of rifling. As the projectile is accelerated up the bore of the barrel there will be no spin imparted to it until it reaches the region in the barrel where the lands are sufficiently intact to engrave the driving bands and so impart spin. If this distance is sufficiently far for the projectile to be moving at a high velocity the sudden engagement of the rifling grooves by the driving bands imposes high shear stresses upon it. If the stresses are sufficiently high the driving bands will be sheared off and the projectile will not achieve gyroscopic stability. An additional problem for aircraft cannon is that the debris from the driving bands may pose a hazard if it is sucked into the aircraft's engine air intakes.

Increase in dispersion may be caused by either wear at the commencement of rifling or at the muzzle. The former may result in the projectile not being fully supported during the critical early stage of its passage up the barrel so that inboard yaw may result. Wear at the muzzle may allow high velocity gas to pass by the projectile, thus disturbing its flight as it leaves the barrel, though such wear is unusual for medium-calibre cannon.

Conclusions

The barrel is the most critical part of a cannon in ensuring that the system meets its performance requirements. Basic design of the rifling profile has been established empirically over a large number of years. Barrel heating and wear are major problems which are ever increasing with the rising muzzle velocities needed to defeat better-protected targets. Medium-calibre cannon often use progressive rifling to reduce the problem of barrel wear but this problem has now reached the stage where barrel life for some cannon systems is measured in only hundreds of rounds.

Annex

Calculation of Accelerating Force on the Driving Bands

The accelerating force, R, acting upon the driving band for a constant twist rifling of angle α can be calculated as follows:

Equating forces acting in the X direction (refer to Figure 4.2):

$$P - R \cdot \sin\alpha - R \cdot \mu \cdot \cos\alpha = m_p \cdot dv/dt \qquad (4.1)$$

where m_p = mass of projectile
dv/dt = acceleration of the projectile in the X direction

Equating forces acting in the y direction:

$$(R \cdot \cos\alpha - R \cdot \mu \cdot \sin\alpha)(D/2) = J \cdot d\omega/dt \qquad (4.2)$$

where $J = m_p \cdot i^2$
= projectile moment of inertia about the X axis
ω = angular velocity
and i = radius of gyration of the projectile.

The angular velocity, ω, and the linear velocity, v, are linked by the rifling twist so that,

$$\omega = (v \cdot \tan\alpha)\, 2/D = (2v/D)(dy/dx) \qquad (4.3)$$

Thus the angular acceleration may be written as:

$$d\omega/dt = 2\,((dv/dt)(dy/dx) + v^2 (d^2y/dx^2))/D \qquad (4.4)$$

Since the angle of twist is small then it may be assumed with little loss of accuracy that $\sin\alpha = 0$, and $\cos\alpha = 1$.

Also, the frictional force, μR, is small compared with the propellant force, P. Thus equation 4.1 may be rewritten as,

$$P = m_p(dv/dt) \qquad (4.5)$$

and equation 4.2 may be rewritten as,

$$R(D/2) = J \cdot d\omega/dt \qquad (4.6)$$

Substituting for J,

$$R(D/2) = m_p \cdot i^2$$

or

$$d\omega/dt = R(D/2)/(m_p \cdot i^2). \qquad (4.7)$$

Substituting equation 4.7 into equation 4.4,

$$R(D/2)/(m_p \cdot i^2) = (2/D)\,[(dv/dt)(dy/dx) + v^2(d^2y/dx^2)] \qquad (4.8)$$

$$\text{or } R = m_p \cdot i^2 (4/D^2)\,[(dv/dt)(dy/dx) + v^2(d^2y/dx^2)]. \qquad (4.9)$$

Substituting for dv/dt from equation 4.5 and simplifying,

$$R = (2i/D)^2\,[(\mathrm{d}y/\mathrm{d}x)P + m_\mathrm{p} \cdot v^2(\mathrm{d}^2y/\mathrm{d}x^2)] \qquad (4.10)$$

From equation 4.10 the angular acceleration force, R, may be found.

Angular acceleration force for constant twist rate

If the angle of twist is constant then the equation for the rifling will be,

$$y = x.\tan\alpha \qquad (4.11)$$

thus

$$\mathrm{d}y/\mathrm{d}x = \tan\alpha \qquad (4.12)$$

$$\mathrm{d}^2y/\mathrm{d}x^2 = 0 \qquad (4.13)$$

Substituting into equation 4.10,

$$R = (2i/D)^2 \cdot P \cdot \tan\alpha \qquad (4.14)$$

Hence for a constant angle of twist the angular accelerating force, R, is proportional to the chamber pressure, P.

5.

Cannon Ammunition Types

Complexity

Medium-calibre cannon have projectiles which are more complex in construction than small arms projectiles. In calibres of 20mm calibre and greater there is sufficient room for mechanical or electronic fuzes and for adequate quantities of high explosive (HE) to ensure a sufficiently large target effect to make the additional complexity worthwhile. This allows for a high degree of flexibility in design so that a projectile can be optimised for a particular target array.

Fuzes

An essential part of a projectile filled with HE is the fuze. The requirements of the fuze are safety in storage, safety during firing, detonation upon impact and self-destruction if the projectile misses the target. These functions are carried out by mechanical fuzes; electronic fuzes are used for proximity or timing functions.

Nose fuzes are typically used in fragmenting HE rounds whilst base fuzes are used in semi-armour piercing HE (SAPHE) rounds. Both nose and base fuzes work in a similar way. Figures 5.1a to 5.1d show the operating principle of the Oerlikon nose fuze type L86A2, which is a typical mechanical fuze that meets the requirements for all of the safety functions. During storage and whilst the shell is travelling up the barrel this fuze has a mechanical block between the firing pin and the detonator *and* the detonator is turned out of alignment with the firing pin. (Most nations require two separate mechanical safety devices for fuzed HE munitions held in storage.) Centrifugal force arms the shell after it has left the barrel. Upon impact the firing pin is released to detonate the HE. If the target is missed the firing pin is released when the spin rate drops below a certain level causing the shell to self-destruct. This type of fuze is very sensitive when armed and will detonate upon impact on even the thin aluminium skin of an aircraft. Fuze sensitivity usually varies with range, because as the spin rate of the projectile decays the centrifugal force holding back the firing pin also decays until self-destruct occurs. Fuze sensitivity must be carefully controlled to ensure that premature detonation cannot occur under such conditions as heavy rain.

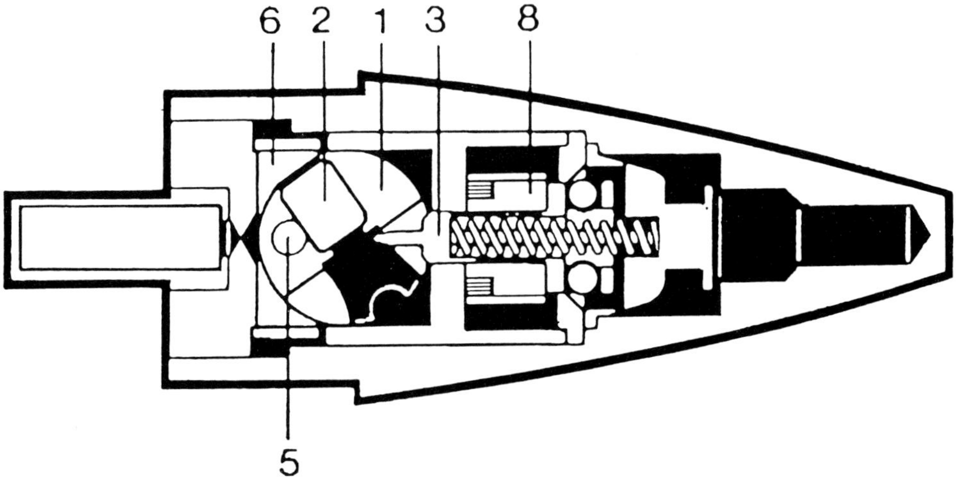

FIG. 5.1a Storage safety. Shutter (1) is locked with the detonator (2) out of line with the firing pin (3) by the security pin (5) and safety lock (6). The firing pin is secured by the split collet (8)

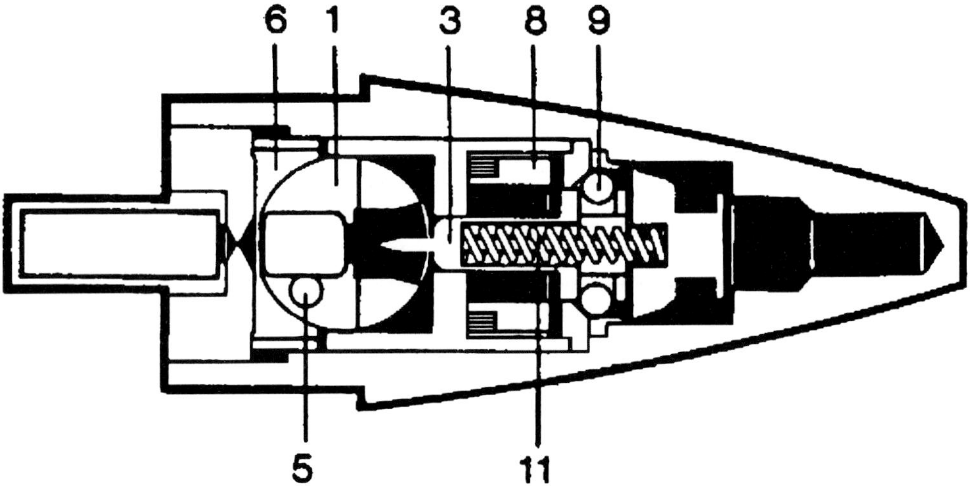

FIG. 5.1b Fuze arming. Centrifugal force pulls back the firing pin (3) by the spheres (9) rolling up the inclined face. The split collet (8) and the safety ring (6) open. The security pin (5) is removed allowing the shutter (1) to rotate. The firing pin is only restrained by the spheres (9)

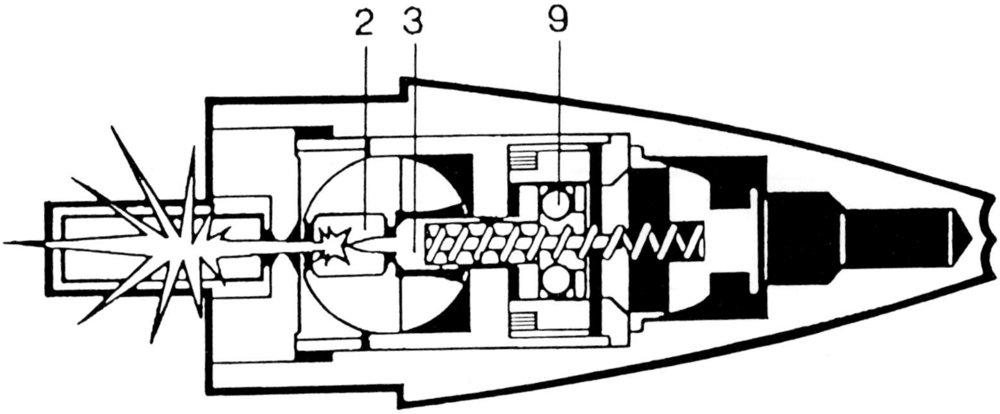

FIG. 5.1c Functioning upon impact. On impact, shock is transmitted to the spheres (9) driving them into their recess and releasing the firing pin (3) to strike the detonator (2)

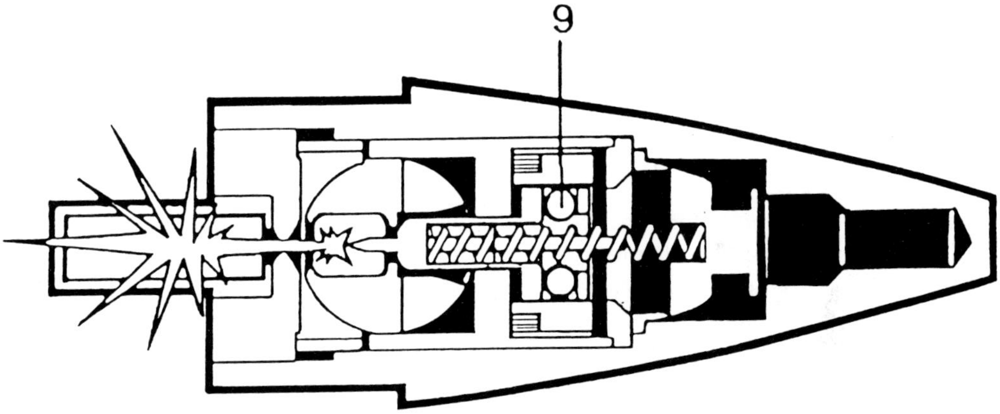

FIG. 5.1d Self-destruction. As the spin rate decays the centrifugal force acting on the spheres (9) diminishes. They are driven back into their recess and the firing pin is released

Types of Projectile

Almost all projectiles have steel bodies to withstand the very high acceleration and centrifugal forces exerted on them as they pass down the barrel. Engraving the body of the projectile directly by the rifling grooves would therefore require a very high engraving force that would damage the rifling lands within the barrel. Short driving bands are therefore fitted close to the base of the projectile. There is a fairly long parallel section of the body of the shell beyond the driving band and towards the nose. This provides a good bearing surface to centre the shell in the lands.

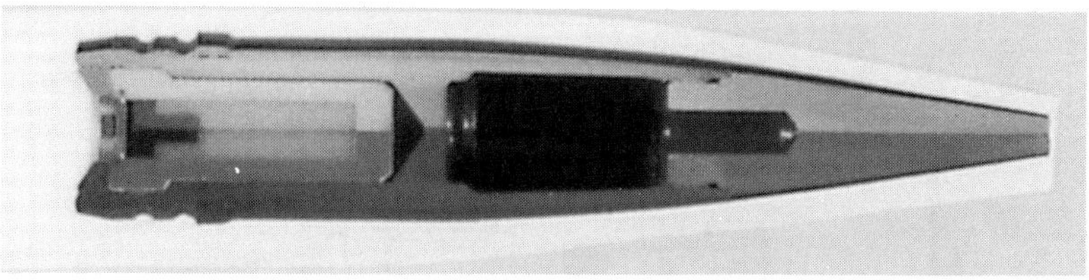

FIG. 5.2 Typical 30mm practice projectile (Courtesy: Oerlikon-Contraves)

Practice Ammunition

Practice ammunition is used extensively for most training purposes. It is inert and so does not require a fuze, although it is usual for the normal tracer element to be fitted. The result is a considerable saving in cost over the more complex operational ammunition, and there is not the problem of unexploded rounds on the range in the event of a fuze failure. For training purposes it is most important that training ammunition is ballistically matched to the operational round for which it is being substituted. To achieve the correct weight distribution it is usual to fit a dummy fuze of the correct weight and to make the body of the projectile hollow. The cavity is filled with inert material of a similar specific gravity to that of HE. Figure 5.2 shows a typical practice projectile.

High Explosive (HE) Shells

High explosive shells have a steel body with a relatively thin wall section. The main body is filled with HE and a fuze is fitted to the nose. A nose fuze is necessary to give sufficient sensitivity against soft targets and must operate when striking the target at shallow angles. When the shell impacts with the target there is a delay before the HE is detonated, thus ensuring that the full explosive effect occurs within the target to cause the maximum damage. This delay is sufficient to allow the shell to travel approximately 1m after impact with the target. The resulting explosion causes the steel casing to fragment. It is these fragments, travelling at velocities up to 1500m/s, which cause damage to the target in addition to the over-pressure blast from the HE. Many HE shells have an incendiary compound incorporated into them which will continue to burn after the HE initiation, thus igniting fuel and other combustible material released by the explosion and resulting fragments. Figure 5.3 shows a cross-section through a typical HE projectile.

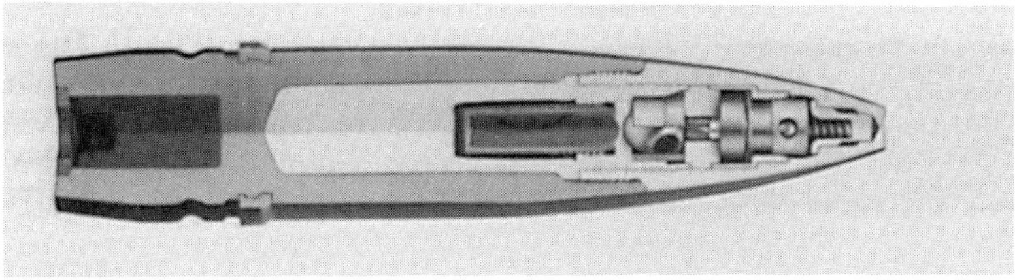

FIG. 5.3 Cross-section through a typical 30mm HE projectile (Courtesy: Oerlikon-Contraves)

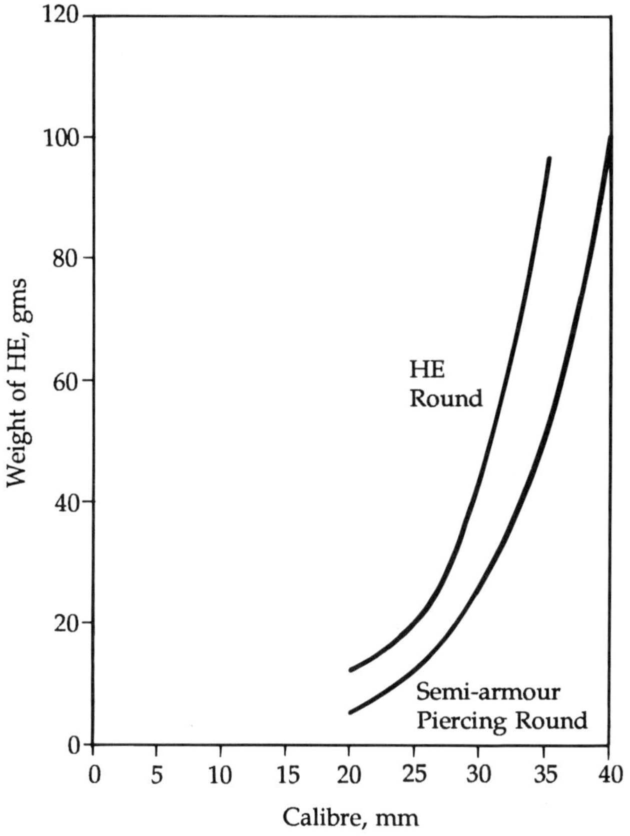

FIG. 5.4 Typical weights of HE in cannon projectiles

Typical muzzle velocities of modern, full-calibre cannon rounds approach 1,100m/s. High velocities are not required to cause damage to the target because it is the HE and the fragments from the shell body that do this. High velocities are required to reduce the flight time of the projectile to a minimum in order to reduce the distance moved by a target between firing and impact. The amount of HE in a 20mm calibre shell is small. As the calibre increases then so does the amount of HE. Figure 5.4 shows the typical weight of HE for different calibres of cannon ammunition. It is generally considered that a minimum of 120g of HE exploding inside a modern aircraft is necessary to ensure a high kill probability. This indicates that a single hit from at least a 35mm shell is required or multiple hits from smaller-calibre weapons.

Semi-Armour Piercing High Explosive (SAPHE) Shells

Semi-armour piercing high explosive shells have thicker walls than an HE round and are usually fitted with a base fuze so that the outer body can be formed round into a solid nose. These are used against hard targets so that a sensitive

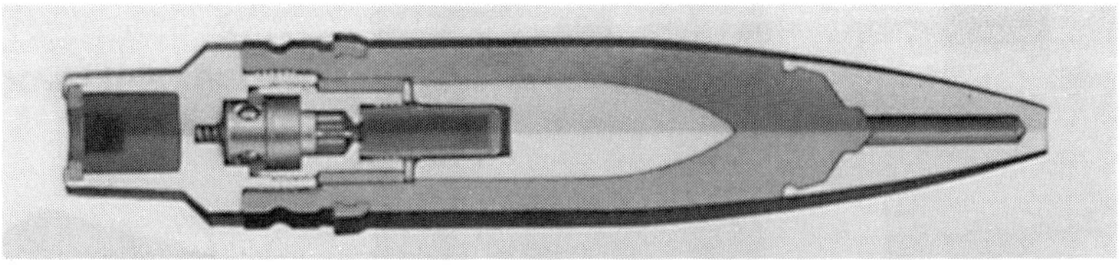

FIG. 5.5 Section through a typical 30mm calibre SAPHE round (Courtesy: Oerlikon-Contraves)

fuze is not required. Again there is a delay between impact and the initiation of the HE so that the shell travels approximately 1 m after impact and before the explosion. This ensures that the HE is initiated inside the target. Damage is caused by the kinetic energy of the shell producing spall, which are fragments from the inside of the armour of the target, and from the steel fragments of the case and the over-pressure blast of the HE. Muzzle velocities and flight times are similar to those of the HE rounds. Figure 5.5 shows a cross-section through a typical SAPHE round.

The amount of space available for HE is much less in a SAPHE round because of the greater wall thickness of the body and because of the solid nose. Figure 5.4 shows the amount of HE contained in the shell body against calibre for typical SAPHE rounds.

Armour-Piercing Discarding Sabot (APDS) Rounds

APDS rounds have a projectile which is smaller than the calibre of the gun. The projectile is supported in a plastic or aluminium segmented sabot. The sabot and the projectile are designed to hold together as they pass down the barrel together. After the projectile has left the barrel the segments of the sabot break apart by a combination of centrifugal force and air pressure. The lightweight unstable sabot segments have a very high aerodynamic drag and typically travel for 50 to 100m down range whilst the sub-calibre core continues to the target. The purpose of this type of construction is to produce a projectile with a high sectional density and high muzzle velocity to increase the thickness of the armour that it will penetrate. Figure 5.6 shows the cross-section through a typical APDS round.

The projectile material is usually tungsten which has a specific gravity of more than twice that of steel. With its low cross-sectional area and high weight, it therefore has a high sectional density.

FIG. 5.6 Cross-section through a typical 30mm calibre APDS round (Courtesy: Oerlikon-Contraves)

This high density ensures that the velocity loss of the projectile with range is low and the armour-penetration performance is high. Since the sabot and projectile together are lighter than a full-calibre projectile, the muzzle velocity is correspondingly higher and can be as high as 1,400m/s. The high muzzle velocity, low loss of velocity with range and high sectional density ensure that the armour-penetration performance is significantly greater than that of a full-calibre round fired from the same weapon.

Armour-Piercing Hard Core (APHC)

Armour-piercing hard core rounds are in many ways similar to APDS rounds. A high-density, sub-calibre core is used, which may be tungsten or depleted uranium. However, a major difference is that the sub-calibre round is supported by a metal jacket which stays with it all the way to the target. Upon impact the jacket remains on the surface of the armour and the hard core continues through the target material. This impact mechanism gives the projectile a high sectional density when it strikes the target and so enhances penetration performance. The weight of the projectile is similar to an HE round and it has the same cross-sectional area, so the external ballistics are similar to those of the HE round. Striking velocity at the target will therefore be lower than that of an APDS round with the subsequent lower penetration performance. Figure 5.7 shows a cross-section through a typical APHC round.

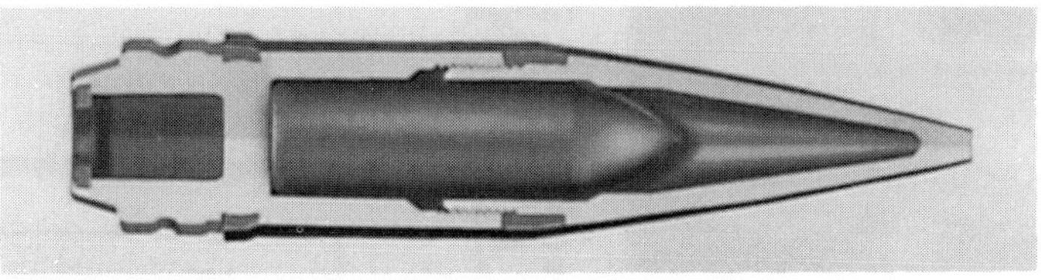

FIG. 5.7 Cross-section through a typical 30mm calibre APHC round (Courtesy: Oerlikon-Contraves)

Armour-Piercing Fin-Stabilised Discarding Sabot (APFSDS)

Fin-stabilised rounds are being developed by many manufacturers but there are few published results on their performance. A typical fin-stabilised round for a 30mm cannon would be 10mm in diameter and 200mm long, made of tungsten with an aluminium ballistic nose cap and aluminium fins. The sabot is usually made of hard plastic or aluminium and the driving band of PTFE. The driving band can be made to slip so that the spin imparted is less than 10 per cent of the full spin rate, which is necessary to prevent the projectile yawing in flight, with the subsequent detrimental effect upon the dispersion. It is difficult to achieve constant performance from slipping driving bands, especially in cannon calibres. Much work is therefore being directed at developing fin-stabilised rounds which can tolerate being spun at the full rate. Oerlikon have developed such a round but it does have a minimum range which is thought to be about 300m. Up to that

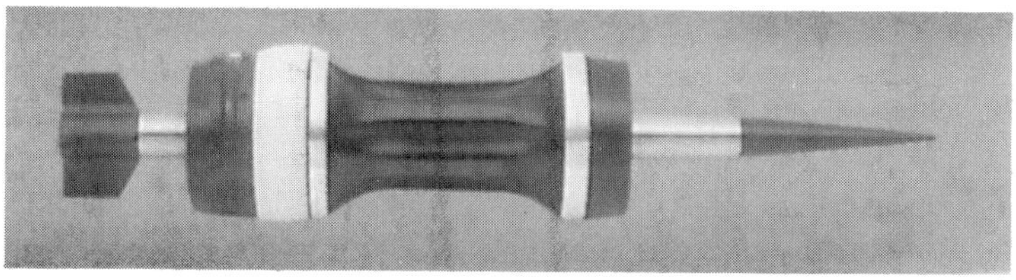

FIG. 5.8 Bofors 40mm calibre fin-stabilised round and sabot (Courtesy: Bofors Weapon Systems)

range the projectile has a high degree of yaw which adversely affects armour penetration.

A problem with fin-stabilised rounds is to release the sabot consistently without influencing the flight of the projectile. Muzzle velocities may be as high as 1,500m/s and the very high sectional density produces a very low velocity loss with range so that armour penetration is high. Figure 5.8 shows the fin-stabilised round and sabot developed by Bofors for their 40mm calibre cannon.

Maximum Effective Range

When considering the maximum effective range of a weapon system the two criteria to consider are the hit chance and the kill chance. For HE rounds the range of the target does not affect the damage that the shell will do to it. Semi-armour-piercing and armour-piercing rounds-however are range dependent for their performance because armour penetration is dependent upon impact velocity, which is affected by range. The prime consideration for HE rounds is the range at which the probability of hitting the target is reduced to an unacceptable level. Although there are many factors affecting hit probability, a dominant one is the flight time of the projectile from the muzzle to the target, especially for aircraft, both fixed wing and helicopters, and soft-skinned targets such as lorries. These targets are likely to be moving so that it is necessary to make an allowance for the distance they have moved during the flight time of the projectile. Additionally, the projectile is influenced by the force of gravity during its flight and an allowance must be made as to how much the projectile will drop from its original path. Wind strength and direction also become significant at longer ranges. Reducing flight times reduces the effects of all of these factors.

It has been found empirically that under normal circumstances a flight time of less than 3s is required to give an acceptable chance of hitting the target. Fig 5.9 shows a typical trajectory diagram for a 25mm calibre HE projectile. It shows the maximum range achieved for different launch angles. Additionally, curves at 1s intervals give the distance that the projectile will travel for different flight times. It can be seen that for a flight time of 3s the projectile will travel 1800m. If the range at which the projectile takes 3s to travel is plotted against weapon calibre then Figure 5.10 is produced. Self-destruct usually occurs after about 8s. The range at which self-destruct for different calibres occurs is also shown in Figure 5.10. This represents the absolute range of the weapon system.

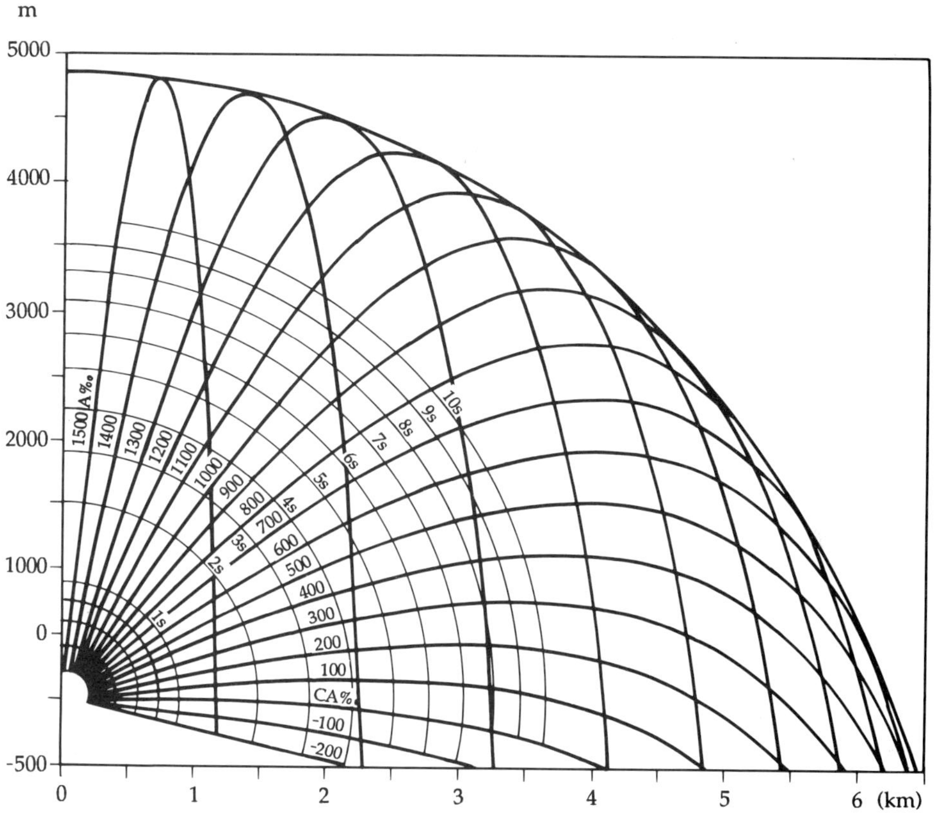

FIG. 5.9 Typical trajectory diagram for a 25mm calibre HE shell (Courtesy: Oerlikon-Contraves)

Figure 5.10 shows clearly that medium-calibre cannon have a very limited range against moving targets if there is to be any reasonable chance of hitting them. The absolute range is also fairly short, which, even for a 35mm calibre weapon, is only 5000m before self-destruct occurs. It can be seen therefore that anti-aircraft and anti-missile cannons are short-range systems compared with missile systems.

Armour-piercing rounds rely upon their ability to penetrate armour to defeat the target. The ability of a round to defeat any armour will depend upon both the armour thickness and the armour material: modern armours could contain steel, ceramic, fibre glass or titanium, for example. It is usual to reduce these materials to a steel equivalent, the steel being rolled homogeneous armour (RHA). The angle at which the round strikes the armour will considerably affect the penetration because the path length of the projectile increases as the slope of the armour from a right angle increases. Armour penetration increases with both the sectional density of the projectile and the velocity at which the projectile strikes the armour. Sectional density increases with calibre. Increases in sectional density also result in a reduction of velocity loss with range so that the striking velocity will be higher, thus further increasing armour penetration.

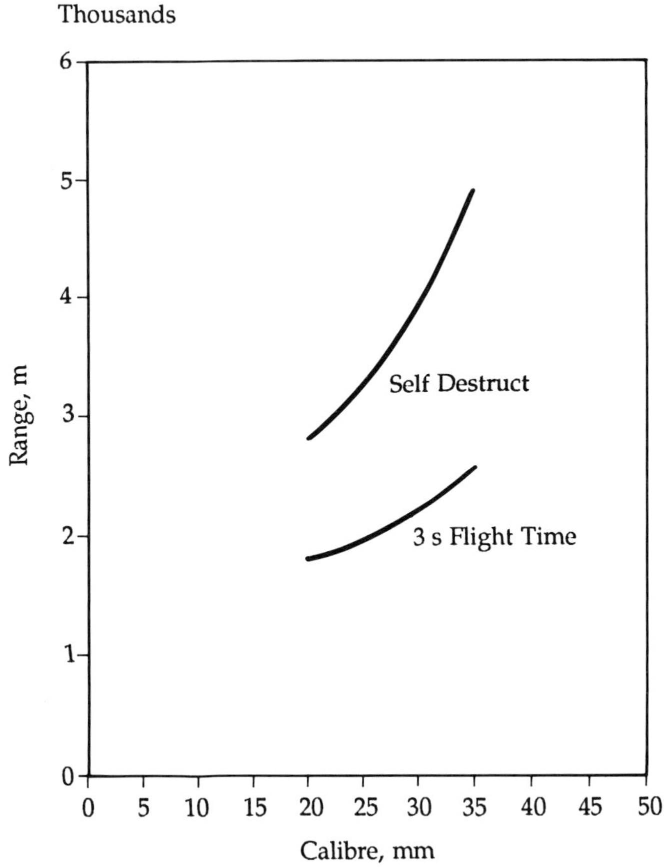

FIG. 5.10 Typical ranges for HE shells of different calibres

Figure 5.11 shows the armour-penetration performance of a typical 20mm calibre SAPHE round for different impact angles and different ranges against RHA. It may be seen that at 1,000m the thickness of armour penetrated is approximately equal to the calibre of the weapon. For a general guide to the performance of different calibres of cannon against RHA it can be taken that for a SAPHE round at 1,000m the armour penetration will be approximately equal to the calibre of the weapon. APDS rounds can penetrate almost twice the thickness of armour that an APSE round is capable of, whilst APFSDS rounds can almost treble the thickness. These improvements are even greater at longer ranges because the sub-calibre rounds lose velocity at a lower rate than the full-calibre rounds and so strike the target at a higher velocity.

The purpose of an armour-piercing round is not just to defeat the armour but also to destroy the target that the armour is protecting. Behind-armour effects are achieved both by fragments being ejected from the back of the armour material and by the residual energy of the projectile after it has passed through the armour. It is usual to specify a greater armour-penetration requirement for a

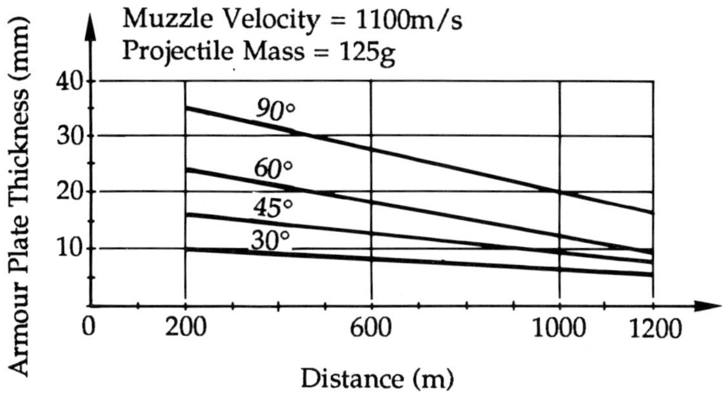

FIG. 5.11 Penetration performance of a typical SAPHE round

specific target to ensure that the projectile has sufficient energy to give the desired behind-armour effect. This over-match is usually in the region of 20 per cent.

The chemical attack of armour with either shaped charges or high explosive squash head (HESH) is well a developed technique in large-calibre weapons. Both of these attack mechanisms are suitable only for use with low-velocity projectiles, which considerably reduces the hit chance. A shaped charge warhead has been developed for the 30mm ADEN round, but as far as it is known there has been no attempt to develop a HESH round for medium-calibre weapons.

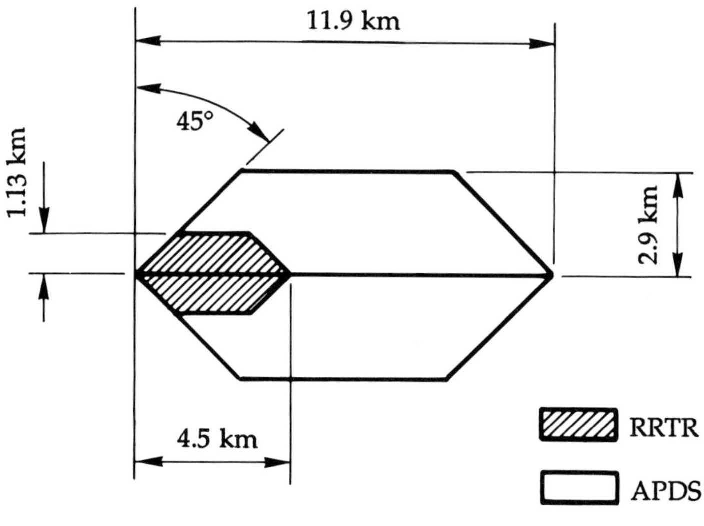

FIG. 5.12 Safety trace for the Royal Ordnance 30mm calibre APDS round and RRTR (Courtesy: Royal Ordnance)

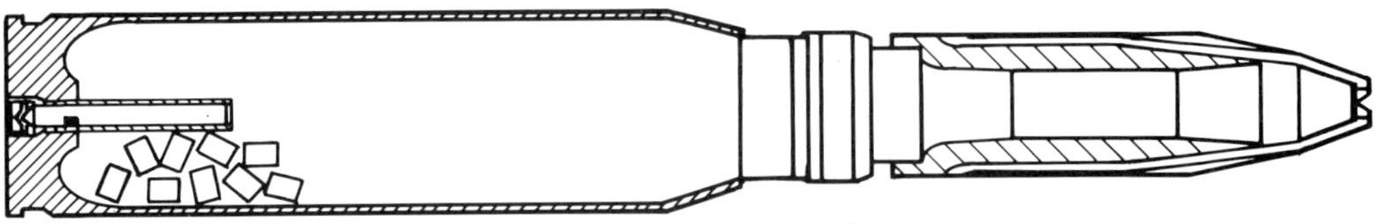

FIG. 5.13 Cross-section through tube reduced range training round (Courtesy: Royal Ordnance)

Reduced Range Training Rounds (RRTR)

APDS and APFSDS rounds are used extensively by many nations to meet the armour-piercing requirements against hard targets. Unfortunately, those very characteristics which enhance armour-penetration performance create problems in training: the low velocity drop with range considerably increases the safety trace of the range on which they are used. A solution to this problem was the development of the reduced range training round (RRTR). This type of projectile is typified by the Royal Ordnance tube round which has been taken into service by the British Army for use in the 30mm RARDEN cannon. It has similar external ballistics to those of the in-service APDS round out to 1,500m. There is then a

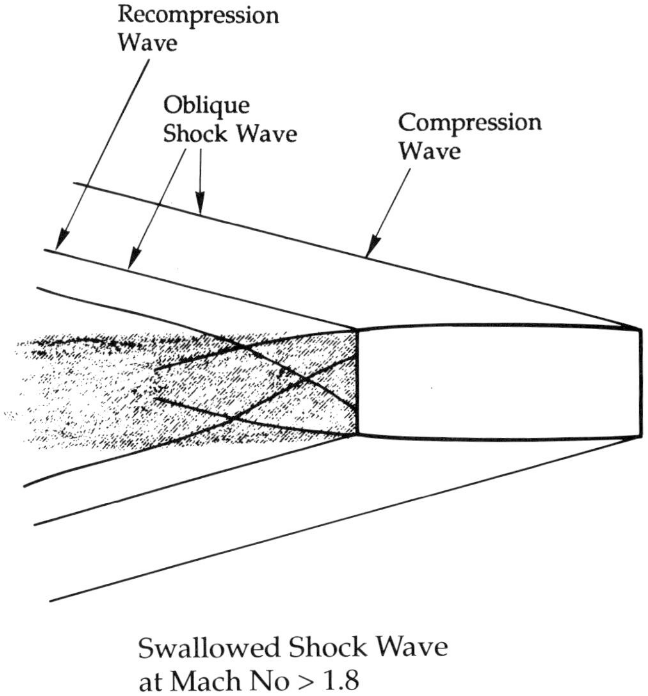

FIG. 5.14 Diagrammatic flow around and through the pipe round at high Mach numbers (Courtesy: Royal Ordnance)

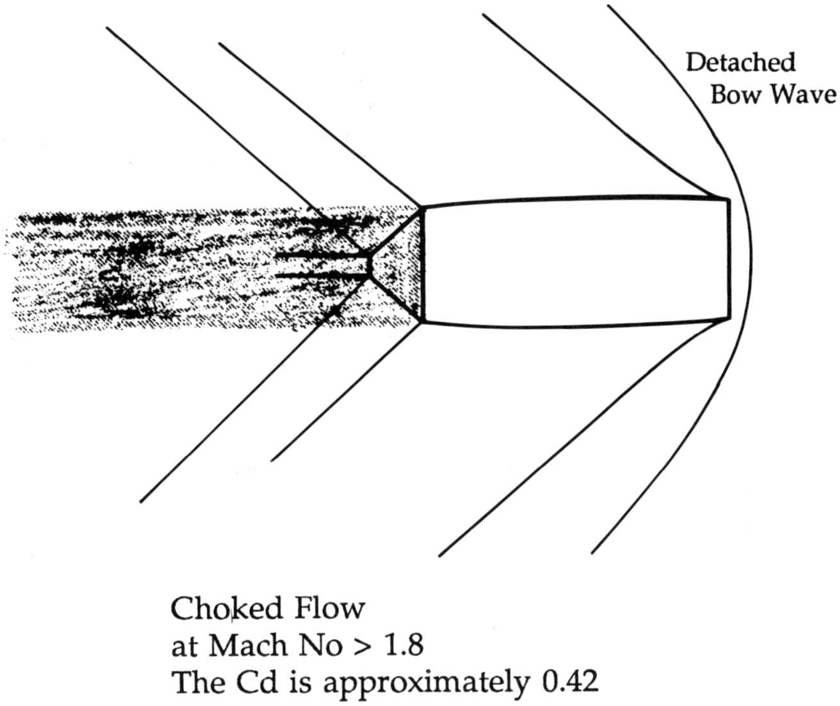

FIG. 5.15 Diagrammatic air flow for choked tube (Courtesy: Royal Ordnance)

dramatic increase in the drag coefficient of the round so that the maximum range is considerably reduced. An additional advantage is the reduced ricochet characteristics, which further reduce the size of the range safety trace. Figure 5.12 shows the safely trace for the 30mm APDS round and the 30mm RRTR. It may be seen that the safety trace for the RRTR is little more than that of a small arms range.

The Royal Ordnance RRTR uses a tube projectile. The complete round is shown in cross-section in Figure 5.13.

The principle behind the tube round is that at high Mach numbers the very small cross-sectional area of the tube gives the projectile a very low drag coefficient so that velocity loss with range is low. Figure 5.14 shows the diagrammatic air flow around and through the tube.

At low Mach numbers the flow through the tube chokes and there is a dramatic increase in drag coefficient with a corresponding increase in velocity loss with range. Figure 5.15 shows the diagrammatic air flow for the choked tube.

It is the sudden increase in velocity drop with range that has a major effect in reducing the maximum range of the round. This increase occurs at about 1,700m and is clearly shown in Figure 5.16.

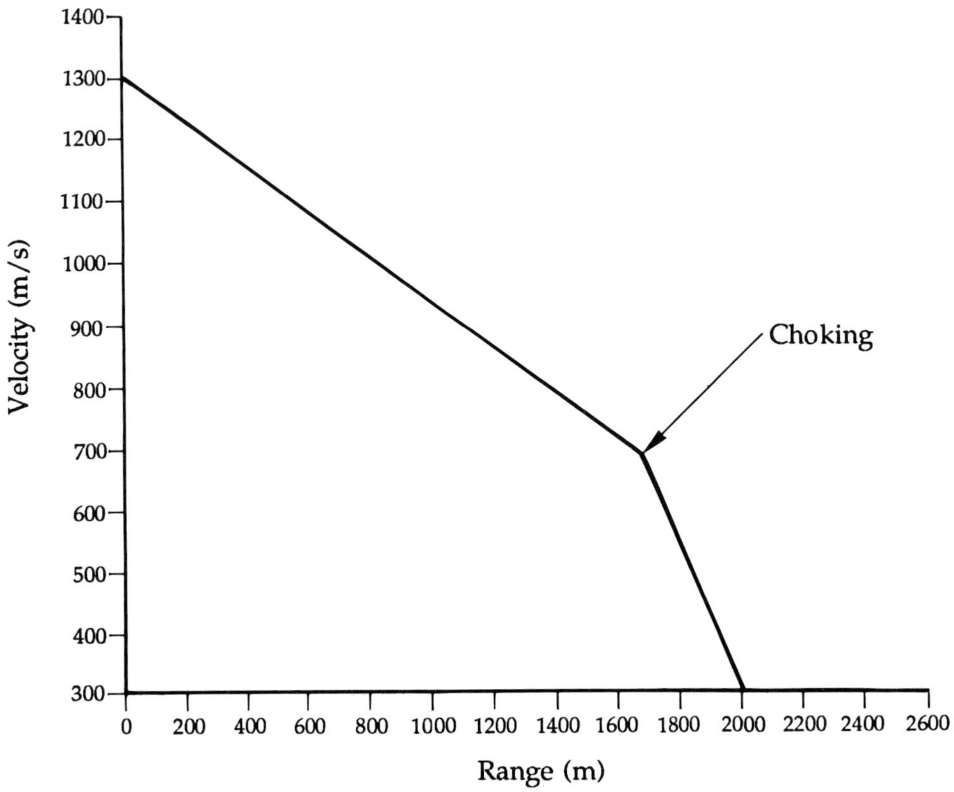

FIG. 5.16 Graph showing change in velocity with range for the RRTR (Courtesy: Royal Ordnance)

Increasing Hit Chance and Range

APDS Rounds

Both range and hit chance may be increased by reducing the flight time of the projectiles. The obvious way is to increase the calibre of the weapon because the greater the calibre then the greater the sectional density of the projectile. This leads to a reduction in velocity loss with range and thus a reduction in the flight time of the projectile. However, another way of increasing sectional density is to use high-density materials such as tungsten or depleted uranium. If the total weight can be reduced then the muzzle velocity can be increased, which is the principle used by the APDS round. Figure 5.17 shows a typical trajectory diagram for a 25mm APDS round. If this is compared with Figure 5.9, which is the trajectory diagram for a 25mm HE round, it can be seen that the flight time for different ranges is considerably reduced. For a 3s flight time the HE shell travels 1,800m, whereas the APDS projectile travels nearly 3,500m.

Many anti-missile cannon systems use conventional APDS ammunition. Not only do the shorter flight times of the projectile increase hit chance (particularly important for such small targets) but most modern missiles have heavily protected warheads and electronics that require the additional armour-piercing capability of APDS rounds to destroy them. Thus the APDS round increases both the hit and the kill chance against this target. Increased range against sea-skimming missiles is particularly important: once the missile is within 500m of its target the destroying of the guidance or the fuzing system is largely academic because it is highly likely that the missile will still hit the target and have sufficient kinetic energy to do considerable damage.

For use against aircraft APDS rounds only result in a kill if the projectile hits a vulnerable area. To increase this chance of hitting a vital part of the target, Oerlikon have developed a frangible round which breaks up upon impact and showers the interior of the target with fragments. Royal Ordnance have developed a similar concept based on their tube round.

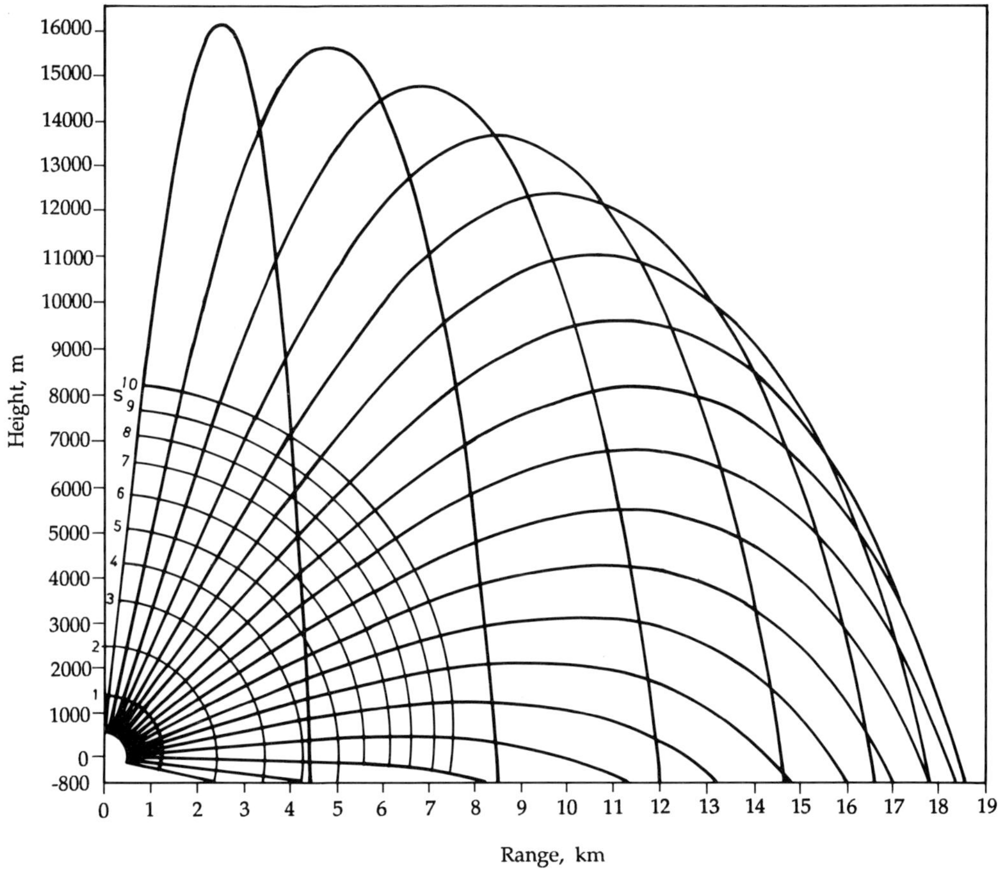

FIG. 5.17 Typical trajectory diagram for a 25mm APDS projectile (Courtesy: Oerlikon-Contraves)

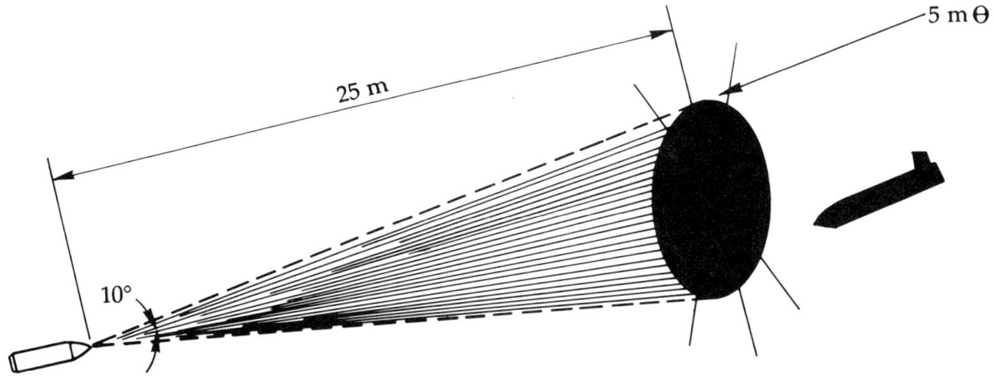

FIG. 5.18 Oerlikon AHEAD-MUNITION anti-aircraft shell concept (Courtesy: Oerlikon-Contraves)

Proximity-Fuzed Ammunition

The purpose of proximity-fuzed ammunition is to produce lethal fragments from the projectile which will hit the target even though the projectile on its own would have missed. There are two different system approaches to this concept. The first explodes a shell in front of the target and produces a cone of fragments which continue on towards it, causing damage to its vulnerable areas. The other approach is to sense when the projectile is within a certain distance of the target and then fragment the shell with HE so that the fragments destroy the target. Oelikon uses the first approach and the principle, known as their AHEAD-MUNITION, is shown in Figure 5.18. The system uses 35mm calibre ammunition and must be used with a fire-control computer and tracking radar. Once the target has been tracked, the computer predicts where it will be when the fire shell is 25m from it. The computer sets the time fuze to explode the shell in front of the target.

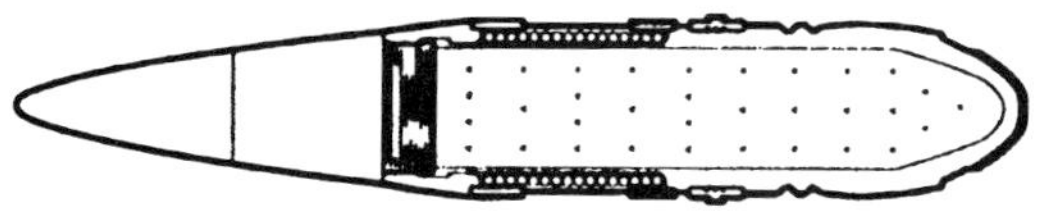

Weight of Shell:	880 g
Weight of Explosive:	120 g
Type of Explosive:	Octol
Number of Spheres:	650
Sphere Diameter:	3 mm
Total Number of Fragments:	>2000
Fragment Velocities:	1000 - 1500 m/s

FIG. 5.19 Details of the Bofors 40mm proximity-fuzed shell (Courtesy: Bofors Weapon Systems)

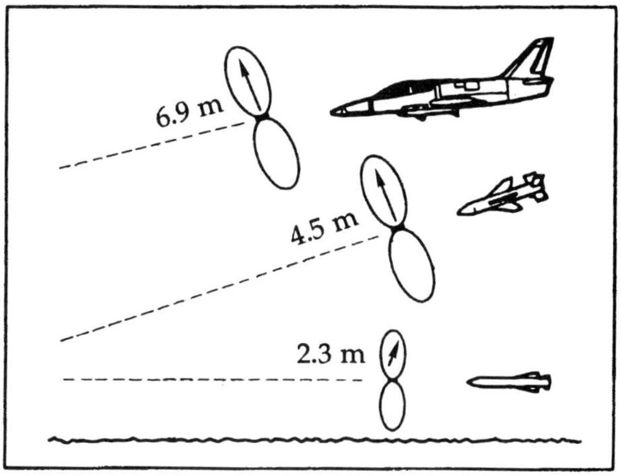

FIG. 5.20 Triggering distances against different targets for the Bofors 40mm proximity-fuzed shell (Courtesy: Bofors Weapon Systems)

True proximity fuzes are used in the Bofors Trinity anti-aircraft cannon system, the projectiles having a radio Doppler fuze. In operation the fuze emits an omnidirectional radio lobe perpendicular to the shell trajectory. When the target appears within a given triggering distance the shell will detonate. The shell also has an impact fuze. Figure 5.19 shows a cross-section through the shell. As well as the fragments from the shell there are 650 3mm diameter tungsten balls to enhance the chance of hit.

It is important that proximity-fuzed shells are not triggered by obstacles near their trajectory on their way to the target. This is a particular problem with sea-skimming missiles. The fuze is therefore set to different triggering distances for different targets. An illustration is given in Figure 5.20

Course-Correcting Ammunition

Considerable effort has been expended to devise course-correcting munitions. The system that has been developed must be used with a fire-control computer and target-tracking radar. The actual projectile is packed around its circumference with HE and has a receiver to receive signals from the fire-control computer. The target is tracked, the computer calculates where the weapon should be aimed to hit the target and the round is fired. The target continues to be tracked by the radar, and the shell on its way to the target is also tracked. The fire-control computer predicts how much, if at all, the shell will miss the target by. At the appropriate time a signal is sent to the shell to detonate a small charge on the required segment on the shell to push its trajectory in the direction of the target. At present this development is being carried out by British Aerospace for the 76mm calibre Oto Melara anti-aircraft cannon, although the technology should be applicable to calibres as small as 30mm.

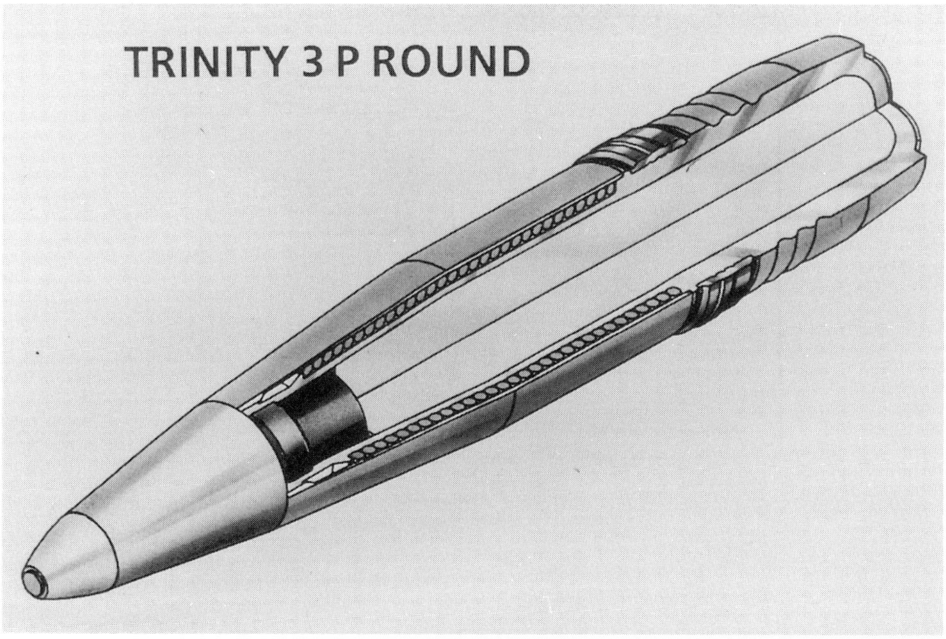

FIG. 5.21 Bofors 40mm calibre 3P multi-purpose round (Courtesy: Bofors Weapon Systems)

Multi-purpose Ammunition

It may be seen that there are a large number of different types of cannon ammunition compared with either small arms or large-calibre ammunition. Each has been optimised for a specific application. However, it is not practicable for a weapon system to be designed to fire a wide range of types selectively. Most systems are designed to select two different ones which are usually an HE and a kinetic energy round. Bofors have developed a multi-purpose round with a programmable electronic fuze which is designed to provide a wide variety of functions; this is based on an HE projectile with tungsten spheres to enhance fragmentation performance. This round is called the 3P multi-purpose round and is shown in section in Figure 5.21.

The 3P round provides all of the functions of a normal HE round, such as safety in storage, arming at launch, initiation on impact and self-destruct. In addition the fuze has a proximity function and a timing function. The proximity function acts in the same way as the dedicated proximity round described earlier. A timing function has been added, not for use against aircraft, but to give an air-burst capability against ground targets. The way that this works is for the gunner to use his laser rangefinder to give the distance to the target. The mode of operation is selected and, as the round is loaded into the chamber, it is programmed for the required time of flight. Rather than being fired direct at the target the round is fired above it so that it gives an air burst above the target, which is showered in fragments. An important capability is for targets to be engaged behind

FIG. 5.22 Bofors 3P multi-purpose round used to give an air burst (Courtesy: Bofors Weapon Systems)

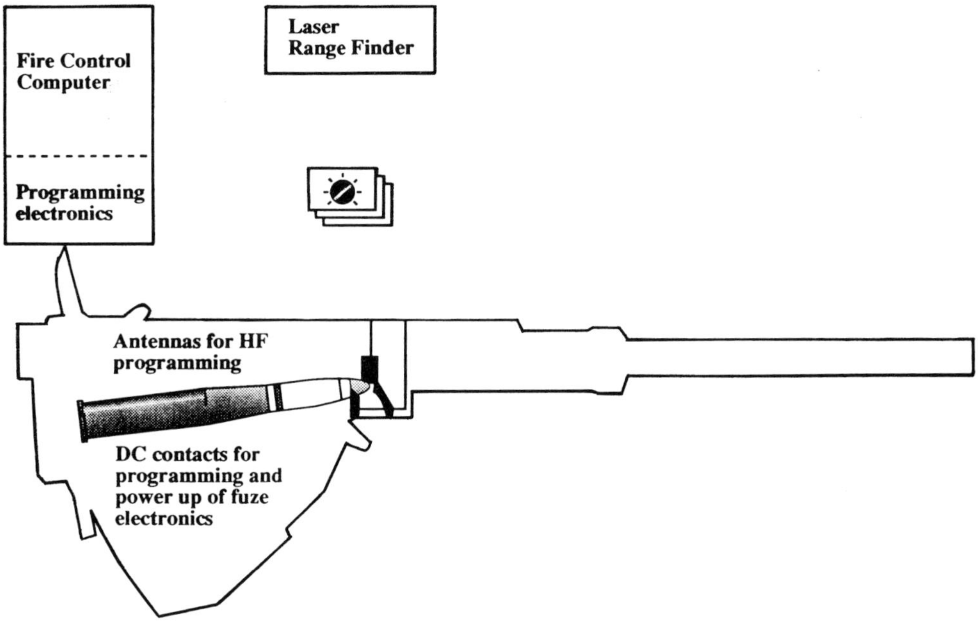

FIG. 5.23 Sequence of events in programming the Bofors multi-purpose round (Courtesy: Bofors Weapon Systems)

a ridge out of direct fire line. The gunner would lase on to the ridge and then add a further distance to where it is estimated that the target is. By firing a burst and with each round programmed for a gradually increasing range a wide area could be covered. Figure 5.22 shows the principle of the air-burst concept.

The sequence of events in the programming of the 3P round is shown in Figure 5.23.

Conclusions

Medium-calibre ammunition takes a wide variety of form, which reflects the considerable number of uses of such systems and the wide range of targets against which it may be used. Sophisticated electronic fuzes have been developed which have added considerably to the effectiveness and the flexibility of the ammunition.

6.

Ammunition Components

Cartridge Case

The cartridge case is a vital component of cannon ammunition, carrying out a number of important functions. It acts as a container for the propellant, a holder for the primer and provides a secure watertight joint between the propellant and the projectile. The cartridge case also provides a seal between the chamber and the breech face against the propellant gases. Most modern cartridge cases are bottle-shaped: their base is considerably larger than the calibre of the weapon. The reasons for this are twofold: to increase the volume of the case without making it unduly long, and to allow the case to be made tapered so that it can be the more easily extracted after firing. An exception is the 30mm calibre ADEN/DEFA aircraft cannon round. This is a relatively low-pressure round that was designed just after the Second World War. The parallel-sided case limits the propellant charge that it can hold, which is reflected in its relatively low muzzle velocity.

To enable the cartridge case to be extracted from the chamber, a rim is fitted. Figure 6.1 shows the three different types of rim found on medium-calibre ammunition. The rimmed round is typified by the 40mm calibre Bofors cartridge. Its basic design can be traced back to the 1930s. The same type of case is still used for the 40mm Bofors gun–although the projectile is one of the most sophisticated encountered in medium-calibre ammunition. The belted case is typified by the 30mm calibre ADEN/DEFA cartridge. Modern conventional ammunition is usually made to the so-called rimless design. This is popular because it is more easy to manufacture than either the rimmed or the belted design. The outside surface of the case does not have any protrusions so that the feeding of the round is simpler; for linked ammunition it can be pushed through the link in the same direction as the feed rather than withdrawn from it in the opposite direction as is necessary with rimmed and belted rounds. However, a disadvantage of the rimless design is that the base of the case must be made substantially thicker to accommodate the extraction groove. This reduces the internal volume of the case and increases both its weight and its cost.

An additional important function of the cartridge case is to reduce the heat transfer from the burning propellant to the barrel chamber. This reduces the problem of cook-off in which the cartridge is fired, not from the external initiation of the firing mechanism, but by the high temperature of the chamber. The

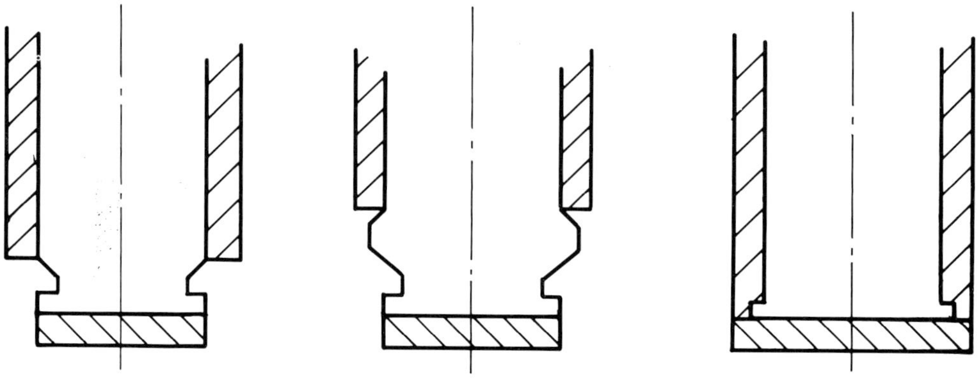

FIG. 6.1 Different cartridge case rims used with medium-calibre ammunition

case will also reduce the heat flow from an already hot chamber to the propellant, reducing further the likelihood of cook-off.

Cartridge Case Material

Cartridge cases have traditionally been made from a copper alloy, usually 70/30 brass. The main reason for this is that brass is a very ductile material which makes the production of cartridge cases, by deep drawing, a simple matter. Brass is also resistant to corrosion and its high ductility ensures that there is good obturation between the cartridge case and the chamber. However, there are several disadvantages associated with brass, the main ones being high cost and low strength. For these reasons, steel is commonly used for the cartridge case material for medium-calibre cannon.

The use of steel enables the case to be made thinner for the same strength, which increases the capacity of the case for the propellant. In addition, case weight is reduced because of the thinner case section and the lower specific gravity of steel (7.85) compared with that of brass (8.5). Another reason for using steel, with its additional strength, is that most applications for cannon require very high rates of fire. For those designs using a single barrel the chamber pressures are still relatively high as the case is extracted from the weapon. During this time the case is unsupported so it must be able to withstand the high hoop stresses set up by the high residual pressure. Whilst brass cases may be able to withstand these stresses, they are often distorted in the process, which may lead in turn to ejection problems.

A major problem associated with steel cases is their lack of corrosion resistance. It is especially so in a marine environment because of the corrosive action of salt spray: steel cases are consequently usually lacquered to protect them. This, however, is only a partial solution because the lacquer may become chipped to expose the unprotected steel. It becomes a particular problem when ammunition is cycled through weapons during exercises and is not actually fired.

As well as strength and density differences between brass and steel another important difference is the Young's modulus. This is the ratio of stress to strain, i.e.,

$$E=\delta/\epsilon \quad (6.1)$$

where E = Young's modulus;
δ = stress
and ϵ = strain.

Young's modulus for steel is approximately 50 per cent greater than for brass and 100 per cent greater than for aluminium. The result is that for the same stress, brass and aluminium alloys will deform more than steel. More importantly, when the stress is released then the brass and aluminium alloys will contract more than steel, thus making extraction of cases made from brass and aluminium easier. Figure 6.2 is the stress/strain diagram for mild steel and brass and shows that, for the same stress, the strain is greater for brass than for steel.

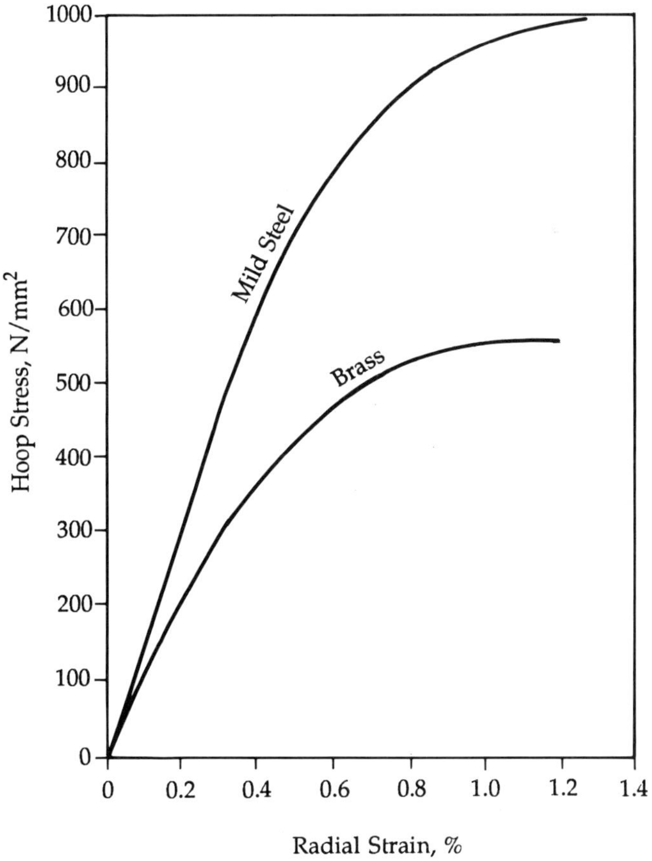

FIG. 6.2 Typical stress/strain relationship for steel and brass

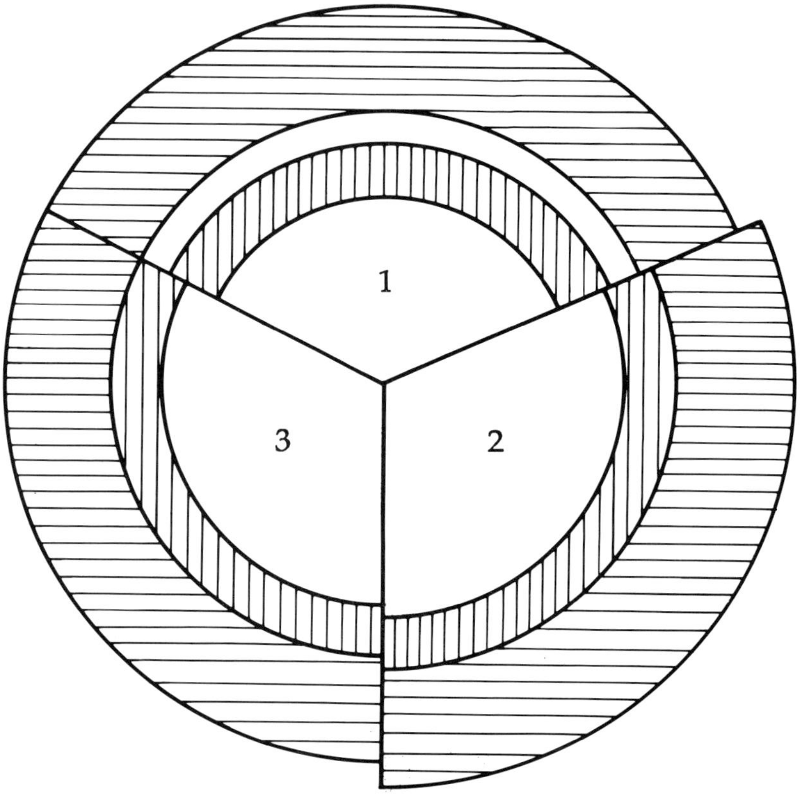

FIG. 6.3 Typical section through the chamber and cartridge case when firing

Figure 6.3 shows a typical cross-section through the chamber and cartridge case of a gun barrel. Section 1 shows the cartridge case and chamber before firing; there is clearance between the case and the chamber. Section 2 shows the chamber and the case during firing; the high pressure has pushed the case to the chamber wall and there has been an expansion of the complete chamber. Section 3 shows the chamber and the case after firing; the chamber has returned to its original size because it was not stressed beyond the elastic limit of the steel. However, because of the very thin walls of the case, the material was stressed beyond the elastic limit and so has not returned to its original size once the pressure was removed.

Figure 6.4 shows this effect for a steel case and for a brass case. Because the case is restrained to expand to the same diameter as the chamber, the hoop strain for both brass and steel will be the same. Both cases will be stretched beyond their elastic limit and will be permanently deformed: when the hoop stress is relieved they will not return to their original dimensions. There will therefore be an interference fit (negative clearance), which in this example will be the same for both cases. There will therefore be a residual compressive stress in both cases, which will make each case grip the chamber and make them difficult to remove. Because of the lower Young's modulus of brass compared to that of steel, the residual hoop stress in the brass case will be lower than in the steel case. The extraction force to remove the steel case will therefore be higher.

In addition to the higher extraction forces required for steel cases, the obturating properties of steel are not as good as those of brass because steel deforms less under pressure. In many instances this is not a problem. However, in turrets which are sealed when firing there may be a problem of toxic gases leaking past the steel case into the turret.

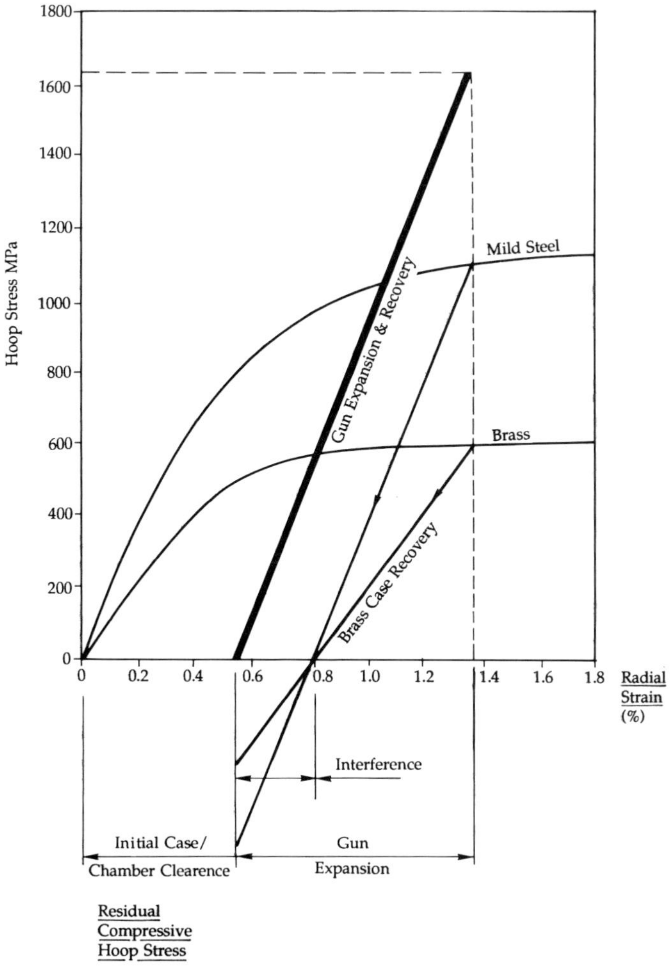

FIG. 6.4 Residual hoop stress in steel and brass cartridge cases

To overcome the problem of high extraction forces associated with steel cases it is common to flute the chamber, this involves the machining of shallow grooves into the chamber parallel with the axis of the barrel from the chamber neck part way towards the breech face. This allows high pressure gas to be directed between the case and the chamber and to push the case away from the chamber walls for part of the length of the case. Obviously, this solution would make the problem of toxic gases leaking into the turret even worse so it is not used for turret-mounted weapons.

Fluted chambers should not be used with brass cases in medium-calibre cannon because brass is so ductile it will actually be formed into the flutes upon firing and make case extraction more difficult.

Aluminium has been used for cartridge cases, but the only known use with medium-calibre cannon is in the 30mm calibre GUA-8 Martin Marietta Gatling cannon. This is an air-to-ground cannon fitted to the A10 ground-attack aircraft. The normal full load of ammunition is 1,170 rounds. The use of aluminium cases reduces the ammunition weight by 272kg for a full payload. Aluminium has a Young's modulus which is lower than that of brass so it does not usually present problems of ejection. It is, however, notch sensitive so that rough ammunition handling may lead to the cases' bursting.

Manufacture of Cartridge Cases

Cartridge cases are usually made from bar stock. To remove the surface defects associated with the rolling process the outer diameter is machined off. Blanks or slugs of exactly the required weight are then sheared off ready for the deep drawing process. The manufacture of cartridge cases is divided into the following five stages:

- Formation of the cavity.
- Drawing the casing wall.
- Pressure moulding.
- Final machining of the case.
- Corrosion protection treatment for steel and aluminium cases.

The basic manufacturing process is similar for all materials. The majority of cases are made of steel and the drawing process for this material requires a specially developed lubrication process based on phosphating between each stage. To produce the cavity for steel cases, the slug (1) in Figure 6.5 is first phosphated with a chemically-applied zinc phosphate coating, which serves as the base for the lubricant used during the subsequent cold working process. The slugs are formed in a press (2) ready for subsequent operations. The slug undergoes work-hardening during each cold working phase and is annealed in a furnace to reinstate the material's ductility. To remove the scale received during the annealing process the slug is pickled, this is followed by phosphating and lubricating. The slug is then impact extruded (3) to form the cavity. After further heat treatments and phosphating the parts are drawn in two (5) or three (6) stages. The final shape of the case is produced in two subsequent operations (7) and (8). Initially a base pressing is carried out to form the base and to strengthen the case by

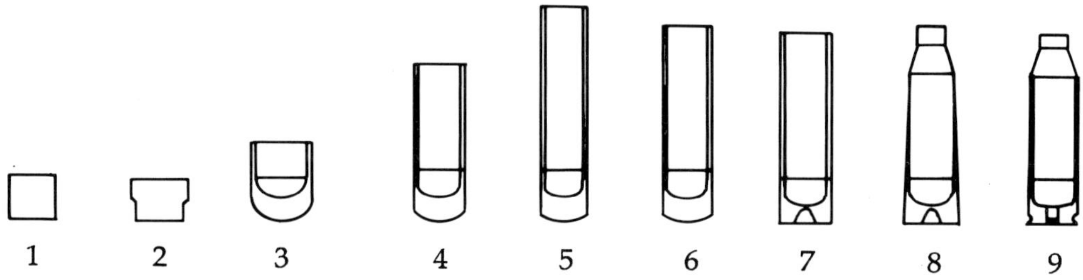

FIG. 6.5 Stages in the production of cartridge cases

work hardening. The subsequent necking operation tapers the neck and forms the shoulder. A machining operation is then carried out to turn the case to the correct length, form the primer hole and produce the extraction groove.

Projectile Bodies

The bodies of practice, HE and SAPHE rounds have an outer case which is referred to as the shell body and fulfils the following functions:

- Contains and secures the explosive.
- Holds the driving band.
- Holds the fuze in place.
- Withstands the gas pressure generated by the propellant.

For HE and SAPHE rounds the shell body is also the source of the fragments to destroy the target.

Projectile bodies are usually turned from solid bar, the amount of material machined away being between 60 and 80 per cent, depending upon the shape of the projectile. Relatively expensive materials are used because good machinability is required combined with high strength. High purity is also required, with freedom from slag inclusions, shrinkage cavities and surface defects. Machining times for boring out the cavity for the HE are high. Cast projectile bodies are rarely used because of the limited quality of the material and the difficulties in achieving optimum and consistent fragment sizes. Forging has proved to be expensive and the large tolerances of the process result in a large amount of machining to produce the finished component.

Primers

The full range of primer types is used with cannon ammunition. The most popular are percussion primers, which are initiated by a firing pin. Their mode of action is the same as for small arms ammunition primers: priming compound, which is sensitive to being struck, is placed between the inner base of the primer cap and an anvil. Ignition occurs when the priming compound is crushed between the cap base and the anvil by the firing pin. Electrical primers, both conducting compound and exploding bridge wire, are often used in cannon ammunition but are usually confined to aircraft ammunition. Conducting composition primers are similar in construction to a percussion primer except that there is no anvil. A firing pin is used which makes contact with the base of the primer cap. However, a voltage is applied rather than an impact. The actual priming compound is mixed with an electrically-conducting compound so that when a voltage is applied to the cap a current flows through it and the conducting compound is heated and the primer is fired. Typical voltages required to fire the primer are between 10 and 18V.

Exploding bridge wire primers work in a similar manner to an electrical circuit protection fuze. A voltage is applied to a wire, creating a current flow that is in excess of the capacity of the wire. Rapid heating of the wire occurs, which causes it to melt, thus firing the priming compound. Typical voltages required for these devices are in the region of 240V dc. To close the circuit a firing pin is placed against the base of the primer cap and a current is then passed through a bridge wire in the base of the primer.

Electrical primers have the disadvantage of requiring an external power source to fire a round. For vehicle-mounted cannon, as most cannon are, this is not usually a problem because the vehicle's electrical system provides the power. An advantage with electrical primers is that they may be used as a means of providing variations in the rate of fire and control over the burst length. A simple electronic circuit incorporated into the primer power supply can delay the firing pulse to reduce the rate of fire below that of the maximum. A simple electronic counter may provide the means of limiting the burst length.

Propellants

Although there is a wide range of propellants currently available, they can usually be classified as single-based, double-based or triple-based. Single-based propellants consist of nitrocellulose (NC) only, whilst double-based propellants consist of differing proportions of nitrocellulose and nitroglycerine (NG). The triple-based propellants contain a high proportion of nitroguandine (Nigu) as well as NC and NG. In addition, there is the recent development in triple-based propellants in which the Nigu is replaced by high-energy nitramine and usually referred to as high-energy nitramine (HEN) propellants. All of these propellants are energetic materials containing the elements carbon, hydrogen, oxygen and nitrogen. The major gaseous products are carbon dioxide and monoxide, hydrogen, water, and nitrogen. NC has a negative oxygen balance of around –35 per cent so that single-based propellants generate a high proportion of carbon monoxide and hydrogen so that their specific energy (energy per unit mass) is relatively low. They also have correspondingly low adiabatic flame temperatures. NG has a positive oxygen balance of +3.5 per cent and when combined with NC in double-based propellants generates greater amounts of carbon dioxide and water so that their specific energies are higher with correspondingly higher adiabatic flame temperatures. Triple-based and HEN propellants generate high amounts of nitrogen and, because of the negative oxygen balance of Nigu and nitramine, the adiabatic flame temperatures tend to be lower. These high-energy propellants are being increasingly used to develop the higher muzzle velocities of modern ammunition.

It is known that one of the most dominant factors determining barrel wear is the adiabatic flame temperature of the propellant. This is the maximum temperature reached when the propellant is burnt in a closed vessel under adiabatic conditions. The adiabatic flame temperature can be determined from the following:

$$T_o = (\Delta U_o/\Sigma Cv) + 298\text{K} \tag{6.1}$$

where T_o = adiabatic flame temperature;
ΔU_o = internal energy of combustion;
and ΣCv = sum of the molar heat capacities

In general, the higher the adiabatic flame temperature then the greater will be the barrel wear rate, because a high temperature leads to a high barrel-surface temperature, which causes a reduction in the physical properties of the steel during the ballistic cycle.

Nitrogen is considerably less erosive than the other gaseous products, and carbon dioxide and water are less erosive than carbon monoxide and hydrogen, with the last being by far the most erosive. Since propellants with the higher adiabatic flame temperatures have gaseous products with the lower erosivity characteristics, the overall erosivity of a propellant is difficult to predict. The only effective way of assessing the wear characteristics of any ammunition is therefore to measure the wear resulting from controlled firing trials.

In general–single-based propellants are the least erosive and double- and triple-based propellants the most erosive. HEN propellants are usually somewhere between. For similar conditions the double-and triple-based propellants can be as much as 20 times more erosive than single-based propellants.

Driving Bands

Cannon ammunition projectile bodies are made of a steel that is of similar hardness to that of the barrel. To engrave the rifling direct into the projectile body would cause considerable damage to the rifling profile in only a few rounds. A driving band of softer material is therefore fitted, the purpose of which is to provide forward obturation of the propellant gases and to transmit the rifling torque to the projectile to provide the necessary spin for gyroscopic stabilisation. The projectile body is made the same diameter as the rifling lands and the driving band is made with the same diameter as the rifling grooves.

Traditionally the driving band material was copper or a copper alloy. These are sufficiently soft to be easily engraved by the rifling and give adequate forward obturation. Modern high-performance cannon ammunition, however, places greater demands on the driving bands, and copper has proved to be insufficiently strong and with too low a melting point for modern propellants. Sintered iron driving bands are now almost exclusively used for full-calibre ammunition.

Sub-calibre ammunition with sabots is usually fitted with PTFE driving bands. Fin-stabilised rounds rely on aerodynamic drag to provide stability in flight, but the barrel through which they are fired is still rifled. If the full spin rate is imparted to them, they yaw badly in flight. Forward obturation is still required so that some form of driving band must still be fitted. It is usual, therefore, for the driving bands to be engraved in the normal manner, but they are usually designed to slip so that the full spin rate is not transmitted to the sabot.

Sabots

Sub-calibre projectiles are used to increase the muzzle velocity and sectional density of the projectile, both of which are desirable properties to increase armour penetration of kinetic rounds. Because the projectile is sub-calibre it must be supported in a sabot during its passage down the bore. Upon emerging from the muzzle the sabot is discarded from the projectile and plays no more part in the ballistics of the projectile. The sabot is made in segments which may number two, three or four. Release from the projectile is achieved by a combination of centrifugal force and aerodynamic drag. It is important that the sabot release is achieved without affecting the flight of the projectile, otherwise there is a detrimental effect upon accuracy. Once separated from the projectile, the sabot petals are highly unstable with a very high aerodynamic drag so that they lose velocity very quickly and usually travel for less than 100m. However, they are highly lethal projectiles in their own right which may endanger anything forward of the muzzle. It is not uncommon for range equipment to be damaged by these flying fragments. Sabots add to the weight of the shot mass. To achieve a higher muzzle velocity than a full-calibre round without resorting to higher chamber pressures the total shot mass must be less than that of the full-calibre round. Sabot weight must therefore be minimised to maximise the weight of the sub-calibre projectile, which is why aluminium and plastics are used.

APDS rounds usually have sabots made of plastic that are made in one piece with carefully designed planes of weakness so that upon leaving the muzzle they break apart, releasing the projectile. An aluminium or steel driving plate is usually incorporated into the base of the sabot. This has two functions. First it keys the projectile to the sabot so that it does not slip within it. Second it transmits the full driving force to the base of the projectile and prevents the sabot being extruded past it.

APFSDS rounds usually have aluminium sabots and are keyed to the projectile by either concentric grooves or by a screw thread running along its length.

Conclusions

The components for medium-calibre cannon ammunition and the materials from which they are made are more varied than for either small arms ammunition or large-calibre ammunition. This once again reflects the wide range of ammunition and the diversity of targets against which types medium-calibre weapons are used.

7.

Recoil Units

The Need for Recoil Units

All conventional gun systems, when fired, are subjected to a force acting upon the point at which they are fixed, known as recoil. This force in small-calibre weapons is small and is usually resisted and controlled by the firer. Below 12.7mm calibre the force is usually insufficiently large to cause problems with the mounting system. For larger-calibre weapon systems the recoil forces are greater. Most 12.7mm calibre weapons have some sort of unit to reduce the recoil force so as to lessen the possibility of damage to the mount and to reduce the movement of the weapon when firing and the detrimental effect that this has on accuracy. Medium-calibre cannon have recoil forces which are sufficiently large to have the potential for causing damage to the weapon and its mount and to cause sufficient movement of the system to have a significant effect upon accuracy. Consequently, recoil units are used to control and to reduce the peak recoil force. They do this by spreading the recoil force throughout the whole firing cycle. Although the peak recoil forces are reduced, the period of time over which they act is increased. When the gun is fired it is set in motion; its direction of movement is opposite to that of the projectile. The recoil unit slows and stops this rearward movement of the gun: the greater the distance over which this is achieved then the smaller will be the recoil force exerted upon the weapon mounting.

Limitations

There is a limit to the movement of the recoil unit because cannon are rapid fire weapons and the recoil unit must return the weapon to its forward position, or nearly so, before the next shot is fired. For high rate of fire weapons there is only a very short length of time to return the weapon to its forward position, so that allowable recoil movement is small. A consequence of this is that recoil forces are high for high rate of fire weapons. The recoil characteristics of a gun can be calculated using Newton's laws of motion. The following simplified approach shows how the recoil velocity of the gun is found, which is then used to find the recoil energy, from which the mean recoil force can then be calculated.

Figure 7.1 shows the diagrammatic representation of a gun firing a projectile.

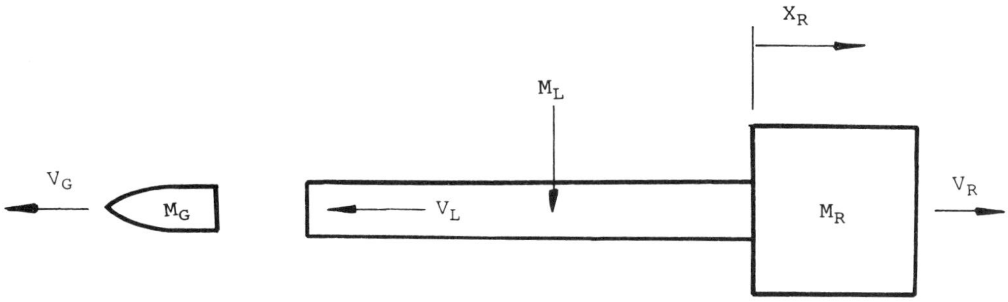

FIG. 7.1 Diagrammatic representation of a gun firing a projectile

The symbols used are:

M_G = mass of projectile
M_R = mass of recoiling parts
M_L = mass of propellant
X_R = recoil distance
V_G = projectile velocity
V_R = recoil velocity
V_L = propellant gas velocity

From Newton's law of conservation of momentum:

$$M_G \cdot V_G + M_L \cdot V_L = M_R \cdot V_R \quad (7.1)$$

When the propellant gases first emerge from the barrel they will be travelling at the same velocity as the projectile. However, the velocity of the following gases will be lower and the velocity of the last particles to emerge will be almost zero. Experience has shown that the mean velocity of the propellant gases may be taken as half the projectile velocity with little loss of accuracy.

Momentum can also be transmitted to the recoiling gun after the projectile has left the barrel by the expanding gases acting upon the end of the muzzle. This additional momentum is known as the after effect. The momentum of the gases plus the after effect, I_T, can be calculated from the following equation:

$$I_T = \beta \cdot M_L . V_G \quad (7.2)$$

where β is a semi-empirical constant calculated from the following equation:

$$\beta = (V_G + 10^6)^{0.5}/V_G \quad (7.3)$$

The conservation of momentum equation may be rewritten as:

$$M_G \cdot V_G + \beta M_L \cdot V_G = M_R \cdot V_R \quad (7.4)$$

or

$$V_R = (M_G V_G + \beta \cdot M_L . V_G)/M_R \quad (7.5)$$

Thus the velocity of the recoiling masses (the gun and its cradle) may be calculated from the recoiling masses, projectile mass, propellant mass and the muzzle velocity. It can be seen that the recoil velocity is increased by increasing the projectile mass, the muzzle velocity or the propellant mass, but recoil velocity is reduced by increasing the mass of the recoiling parts. From the recoil velocity and the mass of the recoiling parts the recoil energy, E, can be calculated:

$$E = (M_R \cdot V_R^2)/2 \tag{7.6}$$

This recoil energy must be absorbed by the recoil system in order to bring the recoiling masses to a halt. The mean recoil force, F, can be calculated from:

$$E = F \cdot X_R \tag{7.7}$$

The mean recoil force is the minimum peak force exerted by the recoil unit on the trunnions of the cannon, because it is assumed that the force is applied constantly over the whole of the recoil stroke.

Rigid Mounts

If the cannon were rigidly mounted, then the recoil force applied to the mounting would be the maximum achievable. This force would be equal to that applied to the base of the cartridge case. The recoil force/time curve will be of a similar form as the pressure/time curve: a fast rise to a peak, which then falls more slowly to zero. The whole time of action of the recoil force will be short, usually between 5 and 10ms.

The peak chamber pressure for the 30mm calibre RARDEN cannon is 400MPa and the inner base area of the cartridge case is 720mm^2. If the barrel were therefore screwed into a proof mounting with no movement during firing then the peak recoil force would be the peak chamber pressure multiplied by the inner base area of the cartridge case, which for the RARDEN cannon, is 288kN. The RARDEN cannon is a long recoil weapon with a recoil movement of 300mm, which is controlled by a pneumatic recuperator and hydraulic cylinder. This reduces the peak recoil force to 13.4kN.

Spring Recoil Units

Spring recoil units are one of the most often used methods of controlling recoil. Springs are used that are compressed by the recoiling gun; the force to compress the spring is that which slows the recoiling gun and its carriage and brings them to rest. Part of the recoil energy is stored in the compressed spring and part is converted to heat by friction as the spring is compressed. The stored energy in the spring is used to return the gun and its cradle to the original forward position. Figure 7.2 shows a typical force/displacement curve for a recoil unit.

Springs have the characteristic that the force needed to compress them is directly proportional to the distance that they are compressed. In the initial, uncompressed state the force to compress them is small and increases to a maximum at maximum compression. It is usual to precompress the spring as shown in Figure 7.2. This reduces the recoil distance and gives less of a variation in recoil force for the unit. Figure 7.2 shows that the spring constant is greater for compressing the spring than it is for extending it. This is caused by the action of friction which always acts to oppose the motion of the spring.

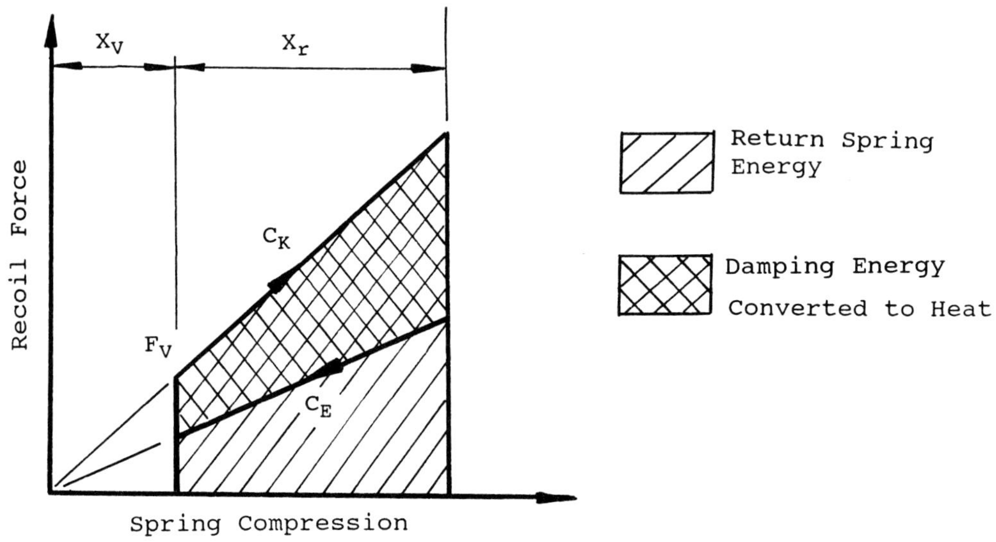

X_V = precompressed displacement
F_V = initial force to precompress the spring
C_E = spring constant during extension
X_r = recoil travel
C_K = spring constant during compression

FIG. 7.2 Force/displacement curve for a typical spring recoil unit

Common coil springs are often used in components in the gun where the movement is relatively large, such as the breech block/barrel return spring in recoil-operated weapons. For recoil units in high rate of fire automatic cannon the spring movement is usually relatively small whilst the compressing force is high, so that a high spring constant is required. Annular ring springs are therefore often used. These consist of a large number of pairs of rings with tapered faces, one fitting inside the other. Figure 7.3 shows a section through one pair of such springs.

When the load is applied, one ring is pushed inside the other and the tapered surfaces exert a radial component of load to compress the internal ring and to expand the outer ring. This results in the axial compression of the ring column, which is the recoil movement of the unit. The actual tapered surfaces slide one within the other when a load is applied to the ring column and this movement is resisted by friction between the two surfaces. Because the rubbing surfaces are large and the surface loading is high the friction forces will also be high. Friction always acts to resist the movement created by the applied force so that there will be a large difference in the spring constant in compression compared with the spring constant in expansion. For annular ring spring units up to 60 per cent of recoil energy is converted into heat by friction so that the heating of the recoil unit will be quite considerable. These types of spring unit are robust and

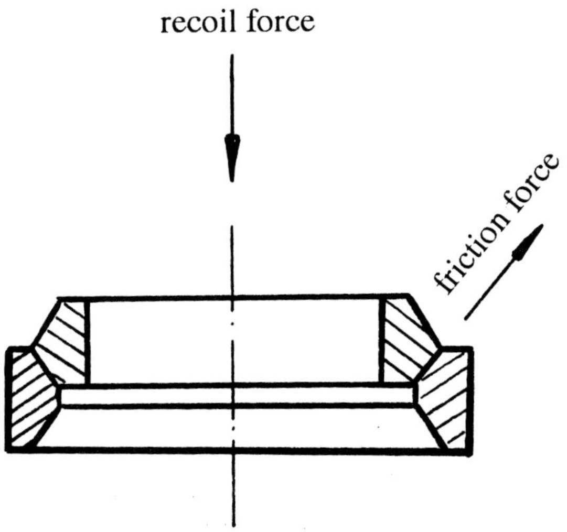

FIG. 7.3 Section through one pair of annular ring springs

reliable and have long service lives. They require little maintenance and are simple to manufacture. Despite relying, in part, for their characteristics on the frictional forces between the tapered surfaces, they are consistent in operation and are little affected by extremes in environmental conditions.

Gas recuperators are sometimes used in place of spring units. They consist of a gas cylinder in which is mounted a piston. A force is exerted on the piston, via a piston rod, by the recoiling gun components, which act in a similar manner to a spring. Preloading of the system is achieved by charging the gas, which is normally air or nitrogen, to the required pressure. When the gas is compressed it follows the normal gas laws so that the force/displacement diagram is different: it is slightly curved. In addition, much less of the recoil energy is converted to heat by friction, so that there is less difference in the spring constant in compression and extension than there is for a mechanical spring unit.

Gas recuperators are compact and light in weight compared with mechanical spring units. They do, however, rely on seals, which require regular attention to ensure that they do not leak. If a leak does occur then specialist equipment is needed to recharge the cylinder.

Friction Brake Units

It is unusual for a recoil unit to rely exclusively on a spring system to control recoil. The recoil energy may be extremely high and reliance upon springs alone may result in excessive stored energy so that the velocity of the gun as it returns to its original position will be so high that there will be the danger of mechanical damage. A supplementary means of dissipating excessive recoil energy is therefore used. One method is to use friction brakes. These are units made in the form of a cylinder fitted with brake shoes acting on the inside surface of the

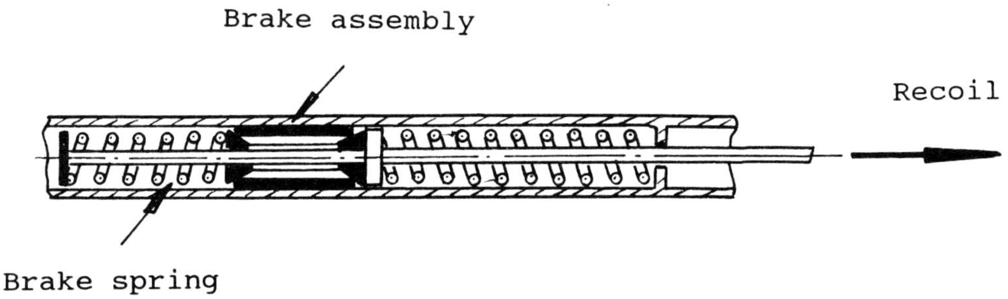

FIG. 7.4 Friction brake with constant braking force

cylinder. The cylinder is fixed to the mount and the brake shoes are fixed, via a rod, to the recoiling gun. Two types of friction brake are used, involving either constant braking force or progressive braking force. A cross-section through a typical constant braking force unit is shown in Figure 7.4.

It may be seen that the braking force can be adjusted by forcing the brake shoes into greater contact with the cylinder bore by lightening the screw unit and forcing out the brake shoes with the tapered collars.

Figure 7.5 shows the principle of the progressive breaking unit.

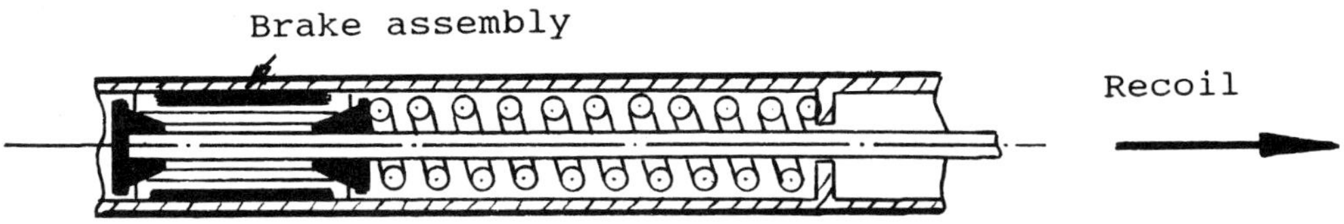

FIG. 7.5 Friction brake with progressive breaking force

It may be seen that the same tapered collar arrangement is used to expand the brake shoes, but the taper is engaged by the movement of the unit. Thus, the greater the recoil movement then the greater will be the braking force.

Friction brakes rely upon correct maintenance and adjustment to function properly. The coefficient of friction between brake pads and cylinder will depend upon the state of the linings and whether the linings are dry or lubricated. All of these factors have reduced the popularity of friction brakes for use in cannon recoil units.

Hydraulic Units

A frequently used way of controlling recoil movement is by the use of a hydraulic unit. This consists of a unit in which is fitted a piston attached to a piston rod that is, in turn, attached to the recoiling gun. The cylinder is rigidly attached to the mount. Figure 7.6 shows a cross-section through a typical hydraulic unit.

The piston can move backwards and forwards in the cylinder, which is filled with hydraulic oil. To allow the piston to move, a transfer passage is drilled

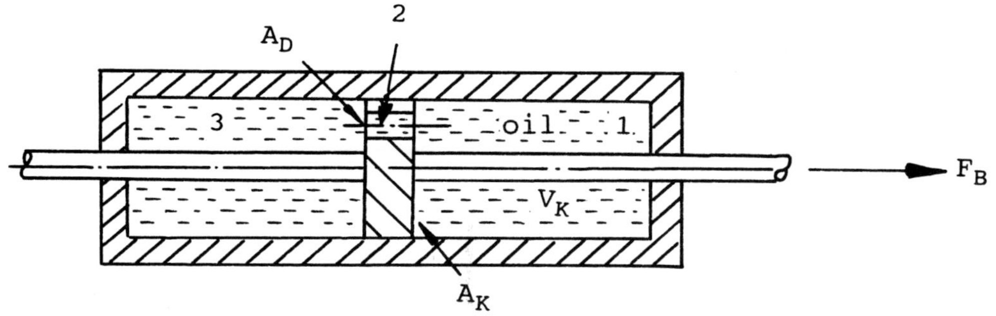

A_D = flow cross-sectional-area
P = hydraulic pressure
A_K = piston cross-sectional-area
ρ = density of hydraulic oil
V_K = piston velocity

FIG. 7.6 Section through a typical hydraulic unit

through it. As the piston moves the oil is transferred from in front of it, through the transfer passage, to the volume behind the piston. Thus there is a higher pressure in front of the piston compared with the back and it is this pressure difference which provides the braking force. This pressure difference is the driving force for the oil flow from the front to the rear of the piston.

Thus the braking force is given by the following (see the legend to Figure 7.6):

$$F_B = A_K(P_1 - P_3) \tag{7.1}$$

The pressure drop between section 1 and section 3 is:

$$P_1 - P_3 = \Delta P_1 + \Delta P_2 \tag{7.2}$$

where $\Delta P_1 = P_1 - P_2$ = pressure drop to accelerate the oil
$\Delta P_2 = P_2 - P_3$ = pressure drop due to fluid friction

The effects of fluid friction are usually small compared with the pressure drop to accelerate the oil and are usually ignored.

Using the Bernoulli equation and ignoring the effects of fluid friction:

$$F_B = 0.5A_K\rho V_D^{\,2} = 0.5A_K\rho(V_K/\mu)^2(A_D/A_K)^2 \tag{7.3}$$

where: V_D = stream velocity in the transfer passage, and
μ = Poisson's ratio of the hydraulic oil

It may be seen that the braking force, F_B, is proportional to the square of the recoil velocity, V_K, so that when the recoil velocity is high the braking force will be high. When the recoil velocity is low the braking force will be low. It can also be seen that the braking force is inversely proportional to the square of the area of the transfer passage through which the oil flows. Braking force can be controlled by controlling this flow area and variable area orifices are often used for this purpose.

Although the fluid friction was ignored because its effect on the braking force is small, the fluid friction is important in converting some of the recoil energy into heat. This results in the hydraulic oil becoming hot.

Hydraulic units are generally light in weight and are very reliable. They do, however, rely on oil seals which must be regularly maintained.

Gun Mount with Fixed Zero Position

When a gun is fired it moves rearward and is returned to the forward position by a spring unit. The characteristic of a gun mount with a fixed zero position is that the recoiling part of the gun returns to its initial position and is at rest before the next round is fired. Figure 7.7 shows the recoil/time diagram for this type of mount.

This type of mount is restricted to use with guns with low rates of fire. Matching the recoil unit to the rate of fire presents few problems, the only requirement being that the cycle time of the recoiling masses should be less than half the interval between shots. When the gun is returned to its original forward position there is an overshoot caused by the gun hitting the front buffer, this sets up an oscillating motion which must be damped before the next shot is fired. This type of mounting is found chiefly in recoil-operated guns and where hydraulic recoil units are being used. A disadvantage of this type of mount is the relatively high recoil forces associated with it because of the need to return the gun to its original position in half the time between shots. The advantage of this mount is the accuracy that it gives from firing from the same position with each shot and the damping of all movement before the next shot is fired.

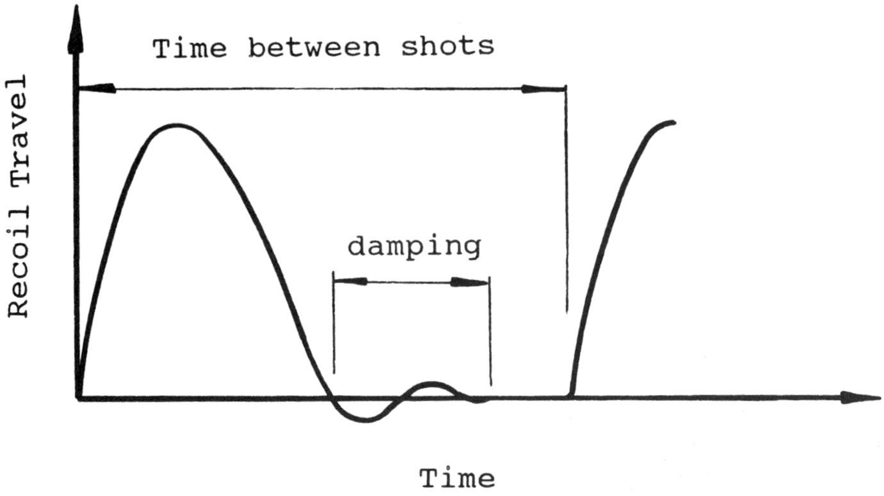

FIG. 7.7 Recoil diagram for gun mounting with fixed zero position

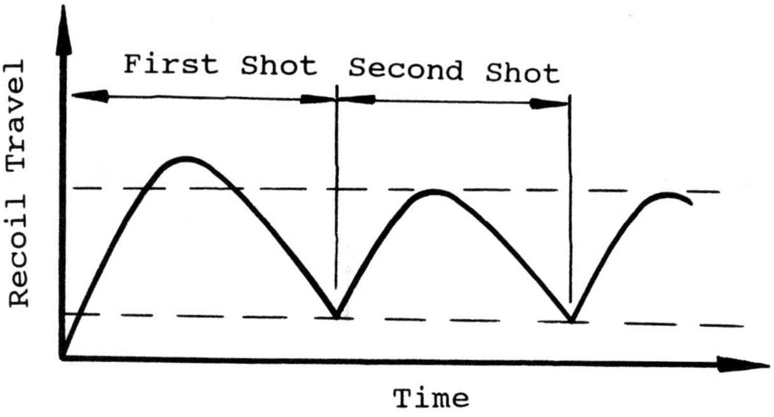

FIG. 7.8 Recoil diagram for a floating mount

Floating Gun Mount

The characteristic of a floating mount is that the second and subsequent rounds are fired when the recoiling parts of the gun are still moving forward by the recoil spring unit. The weapon must always remain on the recoil side of the zero position during firing as this ensures that the recoil forces acting on the mount are always in the same direction and thus that the mount is correspondingly stable. A drawback is the difficulty in arranging the recoil in such a way that the round is always fired during the weapon's forward motion at all rates of fire and angles of elevation. With floating types of mounting, mechanical recoil units are generally used. Figure 7.8 shows the recoil movement/time diagram for a floating gun mount.

The recoil forces are reduced considerably with this type of mount because the movement of the mount can be spread over the same time period as that between shots, and the second and subsequent shots are fired with the gun still moving forward so that it has negative momentum. A disadvantage is that firing the gun whilst it is still moving increases round to round dispersion because it is not being fired from the same position each time relative to the gun mount.

Muzzle Brakes

A method of reducing the recoil force transmitted to the weapon mount is to use a muzzle brake. This is an attachment which deflects the propellant gases rearwards by the use of baffles. The action of the gases impinging upon the baffles creates a force in the same direction as the motion of the projectile. Figure 7.9 shows the basic action of a muzzle brake.

Without a muzzle brake the emerging gases expand in front of the muzzle in a globe. If a baffle is placed beyond the muzzle as shown in Figure 7.9, the gases expand in front of the baffle and are deflected by it. If the baffles are correctly designed the expanding gases are deflected rearwards. This exerts a force on the baffles parallel to the axis of the barrel. This force is caused by the change in

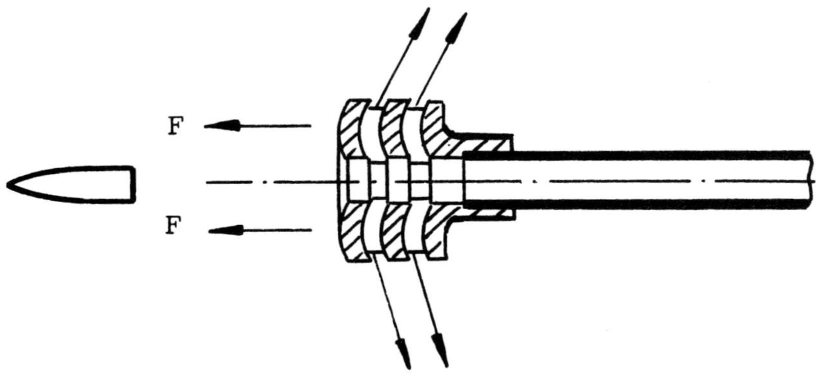

FIG. 7.9 Twin-baffle muzzle brake

momentum of the propellant gases in the direction of the barrel axis and is in the opposite direction to that of the recoil force; thus there is an overall reduction in the force transmitted to the gun mount.

When the gun is fired there is a recoil force transmitted to the mount the instant that the projectile begins to move in the barrel. However, a muzzle brake cannot begin to work until the projectile has left the muzzle, followed by the propellant gases. On a rigid mount this means that a muzzle brake is not very effective because the recoil forces follow a similar profile to the inbore propellant gas pressure; by the time the projectile has emerged from the barrel the peak recoil force has passed and has subsided to a low level. A muzzle brake is most effective when combined with a recoil unit; this is characterised by a progressive increase in the braking force because the muzzle brake acts late in the recoil cycle and will reduce the peak recoil force.

For maximum effectiveness the propellant gases should be directed through 180° back along the axis of the barrel. However, this would be very uncomfortable for anybody standing to the rear of the weapon. Most muzzle brakes are designed to deflect the gases at an angle, usually 45°, to prevent this. Additionally, baffles which point downwards are usually avoided because the deflected gases stir up dust and debris from the ground.

Although up to nine sets of baffles have been used to achieve maximum effectiveness, there is little benefit in using more than three sets, provided they are correctly designed. For medium-calibre cannon, muzzle brakes can reduce the recoil force by up to 30 per cent. It should be noted that baffles should be made so that they are matched at 180° to one another, otherwise there will be an out-of-balance force tending to push on the side of the muzzle, which will have a detrimental effect upon burst fire dispersion.

Conclusions

Recoil forces on medium-calibre cannon are sufficiently large to cause damage to the gun mount and to have a detrimental effect upon accuracy and dispersion unless they are controlled with recoil units. For low rate of fire cannon requiring maximum accuracy fixed-position mounts are used, whereas high rate of fire cannon use floating gun mounts. Mechanical spring units are most often used to spread the recoil force and to return the recoiling gun to its original position, although gas recuperators are sometimes used. Both mechanical friction brakes and hydraulic brakes are used to control the movement of the recoiling parts and absorb the high levels of recoil energy associated with medium-calibre cannon. This may result in considerable heating of the recoil units.

8.

Ammunition Handling

Basic Requirements

An important requirement for all rapid-fire weapons is the provision of the means for feeding ammunition from an external source into the weapon. There may be up to three levels of ammunition storage depending upon the way in which the weapon is used. In the case of aircraft, all of the rounds must be stored so that they are ready to use, because the cannon is sited remotely from the pilot who is not in a position to replenish ammunition from a remote storage area. For anti-aircraft/anti-missile use the engagement times are short with fairly long periods between engagements. Ammunition usage for each engagement is fairly high because of the high rates of fire of the weapon, so that a large number of ready rounds are required. These are replenished between engagements from a remote storage area. Conversely, infantry fighting vehicles (IFVs) usually fire small numbers of rounds and so require relatively few ready rounds. These are replenished between engagements from an easily accessible storage area, whilst the bulk of the ammunition is stored in a remote area of the vehicle. Replenishment of ready ammunition stocks from the different storage areas is usually done manually.

Feeding ready rounds into the weapon may be done by using an on-gun magazine or by using a remotely-positioned magazine with some form of conveyor system linking it to the weapon. Fired case disposal is also very important and in aircraft systems it is usual to have a conveyor system for feeding the empty cases back into the ammunition storage area so that they do not pose a danger to friendly aircraft. Most other cannon eject the fired cases directly to the outside of the vehicle. This overcomes the problem of case storage and, later, case disposal and also prevents toxic gases from the empty cases getting back into the vehicle.

On-Gun Magazines

This type of magazine may be either a fixed integral part of the weapon or a detachable unit which clips directly on to the gun. Examples of the former are the 40mm Bofors L/70 cannon and the Oerlikon 35mm KDE cannon. Both are long-recoil operated weapons that were originally designed as anti-aircraft weapons and have versions that have been developed for mounting in turrets of IFVs. Figure 8.1 shows the Bofors L/70 cannon and Figure 8.2 shows the Oerlikon KDE cannon.

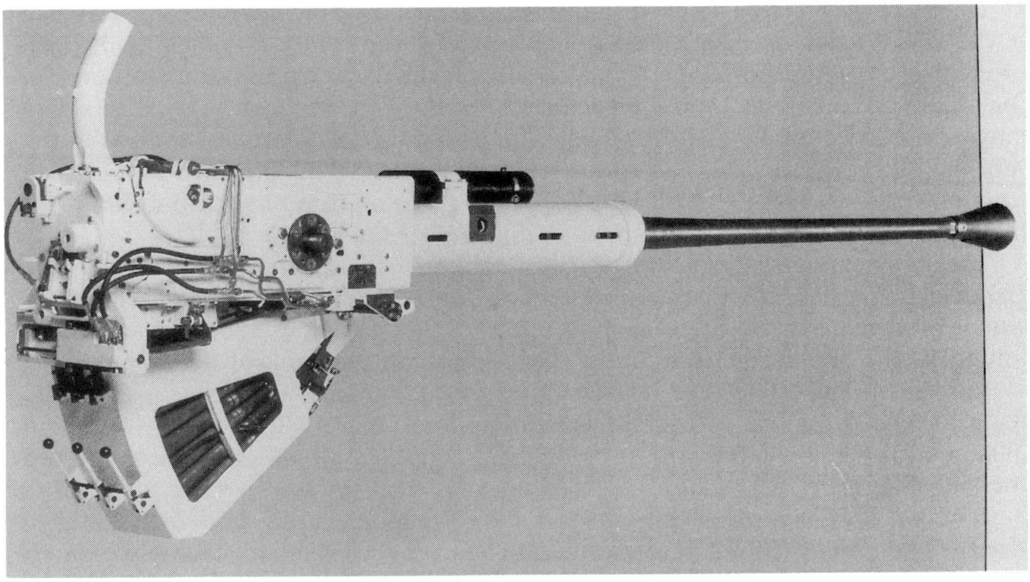

FIG. 8.1 Bofors 40mm calibre L/70 cannon fitted to the CV90
(Courtesy: Bofors Weapon Systems)

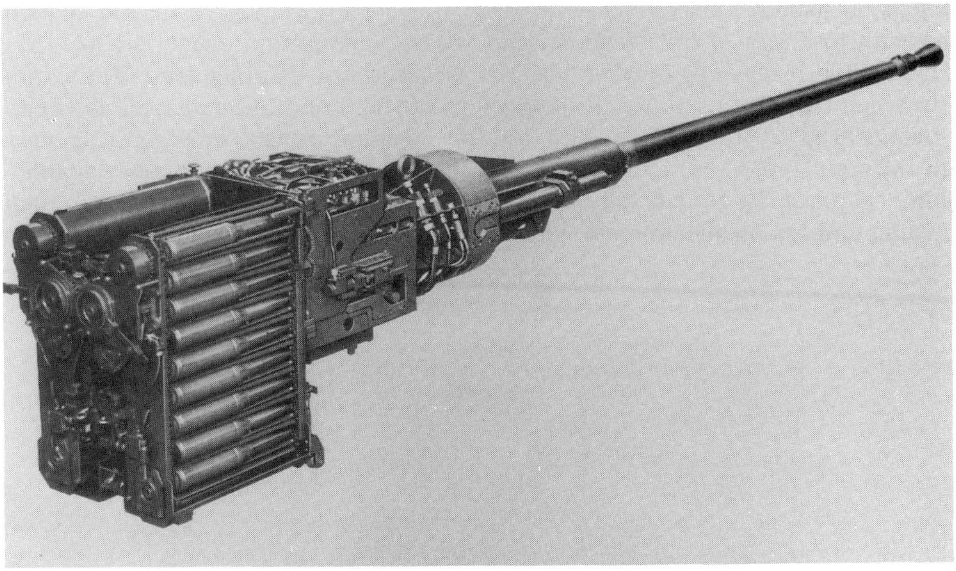

FIG. 8.2 Oerlikon 35mm calibre KDE cannon
(Courtesy: Oerlikon-Contraves)

The Bofors cannon shown in Figure 8.1 is the weapon fitted to the Swedish CV90 IFV. It has a magazine storage capacity of three banks of eight rounds positioned to the rear and beneath the cannon. A different type of ammunition can be stored in each bank, and a lever on the back of the weapon is used to select any of the three banks and thus any of the three types. A carousel system in the bottom of the vehicle holds a further 48 ready rounds which are used to replenish the magazine by hand. A further 168 rounds are stored in ammunition bins, not immediately available, and are held in the rear of the vehicle.

The KDE cannon is mounted in the Oerlikon K80 turret that is fitted to the Japanese M80 IFV. The magazine surrounds the rear of the weapon and holds 17 rounds on each side of the weapon. A different type of ammunition may be stored on each side and selected by the use of a lever on the back of the weapon.

Detachable magazines are usually fitted on to weapons which have originally been designed for use with belt-fed ammunition. Figure 8.3 shows the Oerlikon 20mm KAB cannon fitted with a detachable magazine which clips on to the top of the weapon.

A major advantage of on-gun, magazine-fed weapons is the short feed distance for the live rounds from the magazine to the chamber, which considerably reduces the problems of controlling the movement of the ammunition. Because the magazine moves with the gun, there is no problem of any relative movement of the feed paths between the magazine and the gun as there is with remotely-sited magazines. Cost and complexity are low and in general there are few reliability problems that are encountered with on-gun magazines. However, their storage capacity is usually relatively low so that they are normally restricted to being used with low rate of fire weapons and where ammunition usage is low.

Detachable magazines may be used as an alternative to belt feed. The cannon with which they are used are designed for belt feed and the magazine is capable of being fitted to the cannon with few or no alterations. Their capacity varies with different types and calibres of ammunition but they are usually capable of holding at least 50 rounds. A major advantage is that they may be preloaded with ammunition so that reloading is simply a matter of replacing the magazine.

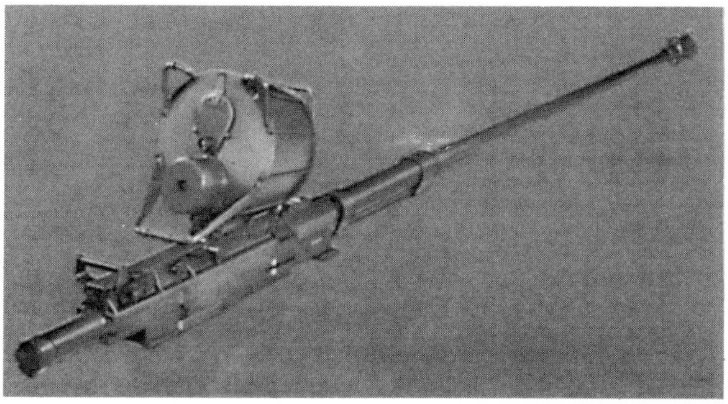

FIG. 8.3 Oerlikon 20mm calibre KAB cannon with detachable magazine (Courtesy: Oerlikon-Contraves)

However, they are, by nature, bulky and occupy a large volume at an inconvenient position on the gun and so cannot be used for turret-mounted weapons.

On-gun magazines increase the volume and weight of the weapon. This additional weight and volume may be critical because the weapons using this type of feed system are often fitted into the turrets of AFVs where these factors are extremely important. An additional problem is that the volume swept inside the turret by the gun in elevation and depression is increased. Provision must also be made for access for loading and unloading the magazine as well as loading and unloading the cannon. However, there is a considerable saving in volume from the elimination of the need for any conveyor system as used in remote magazines. This volume saving may be considerable, especially where a dual conveyor feed is used for firing two different types of ammunition.

The additional weight of the magazine and the ammunition to the back of the weapon increases the loads placed on the trunnions and the power requirement of the elevating gear. The weapons using on-gun magazines are usually of long-recoil operation which already have the characteristic of a varying centre of gravity, when firing, which is compounded by the change in the centre of gravity of the system caused by using the ammunition in the magazine.

The biggest drawback of on-gun magazines is their limited ammunition capacity and the time taken to reload them. Reducing the system volume and weight by reducing the number of rounds in the magazine increases the problem. This is typified by the 30mm calibre RARDEN cannon which was designed for minimum recoil force, minimum weight and minimum inboard volume. To achieve

FIG. 8.4 Three-round clip ready to be loaded into the RARDEN cannon

this, long-recoil operation was used and the feed was limited to a three-round clip hand-fed into the back of the weapon. Figure 8.4 shows the rear of the RARDEN cannon with a three-round clip ready to be loaded. Because there is no external feed system and the magazine capacity is only three rounds the weapon is very compact. The clip system also makes the weapon quick to reload. However, this method of feeding the weapon requires two men to fire it, the gunner and the commander/loader. This has operational implications as it ties two men to the gun and they are not available for other roles.

Advantages of on-gun magazines are:

- Short feed distance.
- Reduced overall volume occupied by feed system.
- Low cost.
- Reliability.
- Simple selection of different natures of ammunition.
- External power not required.

Disadvantages of on-gun magazines are:

- Increases weight and volume of the weapon.
- Limits the number of ready rounds.
- Change in centre of gravity of weapon with ammunition load.

Remote Magazines

Remote magazines store ammunition away from the weapon and transfer it from the storage point to the cannon by some form of conveyor system. Almost every cannon application with this type of feed has a system designed to meet its individual requirements, consequently the different types of design are considerable. However, these designs usually follow a small number of patterns, which will be considered here.

Belt Feed

The simplest form of remote magazine uses linked ammunition. On modern cannon these links are of the disintegrating type. Older cannon used continuous belts, especially aircraft cannon where the belt was used for replacing the fired case for safe storage. Although the ammunition is linked together to form a belt, it is usual to use flexible ducting to guide and control the rounds from the ammunition box to the weapon. Flexibility is required because the weapon will be traversed and elevated in use so that the feed position will vary with respect to the ammunition bin, which is usually fixed. For short feed distances the power from the weapon is used to power the belt. For long distances and for lift heights over 1m it is usual to have auxiliary power to help feed the ammunition.

For cannon with dual ammunition feed it is necessary to have two separate magazines with two separate, flexible feed ducts. The 30mm calibre M230 chain gun has been adapted for many uses, including mounting on ships, aircraft and light vehicles. Figure 8.5 shows the M230 chain gun mounted on a light vehicle

FIG. 8.5 Flexible feed ducting used with the M230 chain gun (Courtesy: McDonnell Douglas)

and firing linked ammunition that is controlled by a flexible feed chute between the ammunition box and the weapon.

The greatest advantage of remote magazines is that they are able to provide a large number of rounds without replenishment, which is essential for weapons with a high rate of fire or where large numbers of rounds are fired per engagement. Linked ammunition is particularly convenient because preloading into the remote magazine is not necessary: the ammunition box is used as the magazine. Linked ammunition does, however, have the problem that the links must be disposed of. Most systems collect them in the turret and dispose of them manually when it is convenient. Loose links in the turret may be a hazard and if they are allowed to build up they can cause the weapon to malfunction.

Belt feed with rimmed or belted ammunition is difficult to achieve because the round cannot be pushed through the link by the forward-moving breech bolt. To feed such ammunition from a belt requires the round to be pulled rearwards from it and then fed downwards on to the bolt face, which adds to the complexity of the feed system.

Advantages of linked-belt feed systems are:

- Capable of providing a large number of ready rounds.
- Reloading of ready rounds is simple and quick.
- The conveyor system is mechanically simple.

Disadvantages of linked-belt feed systems are:

- Two feed paths are required for two ammunition types.
- Feed paths occupy a large volume.
- Provision must be made for changes in feed position with gun elevation.
- External power required for long belt lengths.

Linkless Conveyors

Linkless conveyors are used with aircraft cannon system applications and many anti-aircraft systems. Many aircraft cannon are fitted with linkless conveyors because these weapons have very high rates of fire that require large-calibre rounds to be fed into the weapon at very high rates. Multi-barrelled Gatling cannon have highly developed feed systems that transport live rounds from a remote magazine to the gun and remove the fired cases from the weapon and feed them back into the remote magazine. Figure 8.6 shows the schematic arrangement for the feed mechanism used with the GUA-8 30mm Gatling cannon fitted to the A10 ground-attack aircraft and the ship-borne Goalkeeper anti-missile system.

The major parts of the feed system are:

a *Drum assembly*. Ammunition is stored in the cylindrical drum assembly. Live rounds are stored radially pointing inwards in a helix in the inner drum. This inner drum rotates and the live rounds move along the helix to the exit unit.

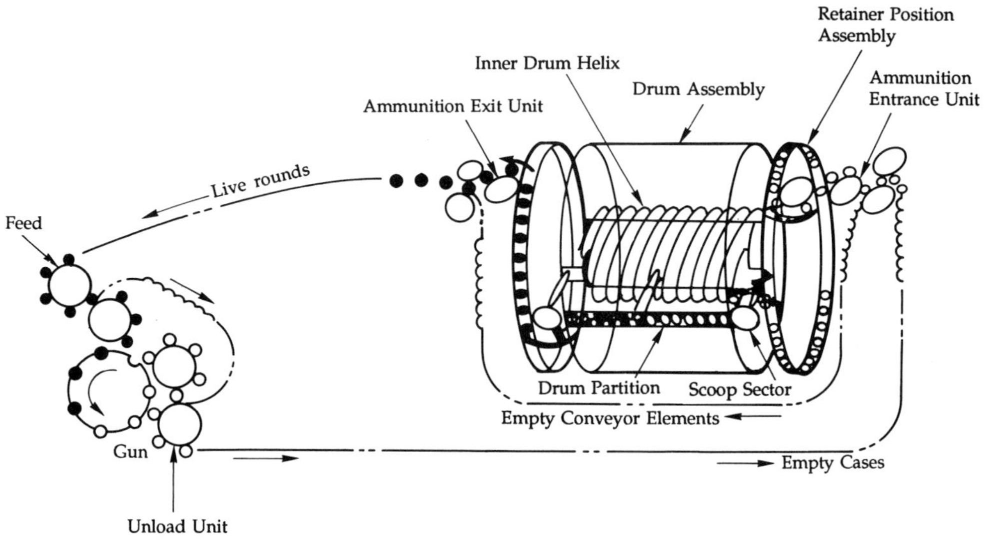

FIG. 8.6 Linkless conveyor system used with Gatling cannon

b *Ammunition exit unit.* Ammunition is removed from the inner drum by the ammunition exit unit, which takes the live rounds and inserts them into an endless conveyor system. The conveyor transports the live rounds to the weapon.

c *Transfer unit.* The live rounds are removed from the conveyor system by the transfer unit, which feeds them into the cannon. Empty cases are removed from the cannon and reinserted back into the conveyor system, which transports them back into the cylindrical drum unit.

d *Ammunition entrance unit.* The empty cases are removed from the conveyor system by the ammunition entrance unit, which transfers them back into the cylindrical drum unit where they are stored in the space vacated by the live rounds.

Many of the large-calibre, gas-operated anti-aircraft systems have linkless feed systems. The mechanical design of the actual storage area and the transfer mechanism may be different but they perform the same function. They are also often simpler because there is not the need to return the empty cases to the magazine. A major difference between the linkless systems used with gas-operated weapons and those used with Gatling cannon is that the former have provision for storing and selecting different types of ammunition. Although the complete feed system is linkless, the ammunition is often linked for initial loading into the storage area. An externally powered unit takes linked ammunition from the storage boxes and feeds the rounds into the drum assembly, delinking them as it does so.

Much development work is also being undertaken to develop linkless feed systems for IFVs, especially with the trend in ever-increasing calibres for these weapons.

Advantages of linkless feed systems are:

- Give complete control over the feeding of the round.
- Capable of feeding weapons with a very high rate of fire.
- Capable of controlling and storing the empty cases.

Disadvantages of linkless feed systems are:

- Require external power.
- The systems are large, complex and expensive.
- Reloading the systems is time consuming.

Conclusions

An essential component of all cannon systems is the feed mechanism that takes ammunition from a storage source and feeds it into the weapon. For low rate of fire weapons that fire few rounds in an engagement the feed system can often use on-gun magazines with great effect. For weapons with higher rates of fire and where the number of rounds fired is large, the complexity of the feed system increases. Very high rate of fire weapons require sophisticated systems which are both complex and bulky.

9.
Cannon Weapon Systems

The Need

A cannon system, of whatever calibre and configuration, is a large and heavy weapon that, unlike a small arm, cannot function without the aid of other components. The actual weapon must be mounted on a structure to give it stability and provision must be made for aligning it with the target. Actual alignment of the cannon with the target is only required for stationary targets; for moving targets the cannon must be aligned with a point in space so that the projectile intercepts the target. Most of the cost and complexity of a cannon system results from the provision of different means to increase hit probability. This may be in the form of tracking systems, computer systems to predict the path of the target and where to aim the weapon, or for systems to feed very high rate of fire weapons to saturate the target with the required number of projectiles. Some systems will be provided with all of these. The final configuration will depend upon the role for which the weapon is to be used. There are almost as many different cannon systems as there are cannon. The purpose of this chapter is to describe the modern systems developed for a range of applications and to use examples to illustrate the different developments.

Aircraft Cannon Systems

Cannon fitted to fixed wing aircraft vary from 20 to 30mm calibre. The Americans favour the externally-powered, multi-barrelled Gatling cannon, whereas the Europeans tend to favour gas-operated, single-barrelled, revolver cannon. These revolver cannon are often used in pairs to give similar rates of fire to a single Gatling cannon. However, the actual systems for the two weapons are very similar and consist of the cannon and the feed system. The cannon system may be incorporated into the body of the aircraft or may be fitted into externally-mounted pods which are easily removable. Externally-mounted pods offer the greatest flexibility in weapon selection and for certain aircraft, such as the F4 Phantom, it is possible to fit three pods fitted with Gatling cannon to give a combined rate of fire of 18,000 rounds per minute.

An example of a modern, externally-mounted, fixed-wing aircraft cannon system is the 25mm Gatling cannon system produced by Martin Marietta for the AV-8B vertical short take-off and landing (VSTOL) aircraft. The system is made up of two pods which are fitted to the bottom port and starboard sides of the fuselage.

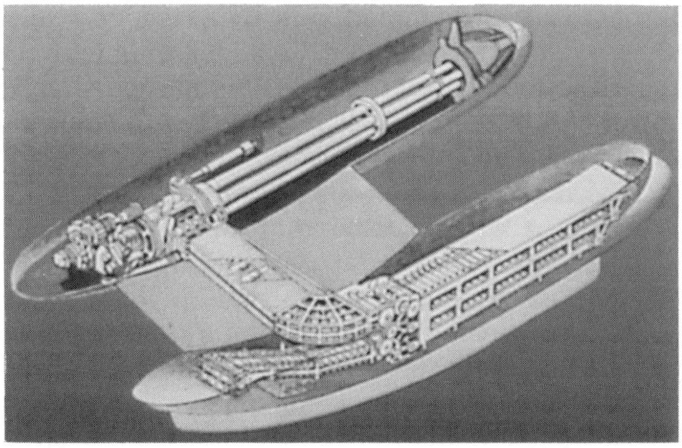

Fig. 9.1 Martin Marietta AV-8B 25mm calibre gun system (Courtesy: Martin Marietta)

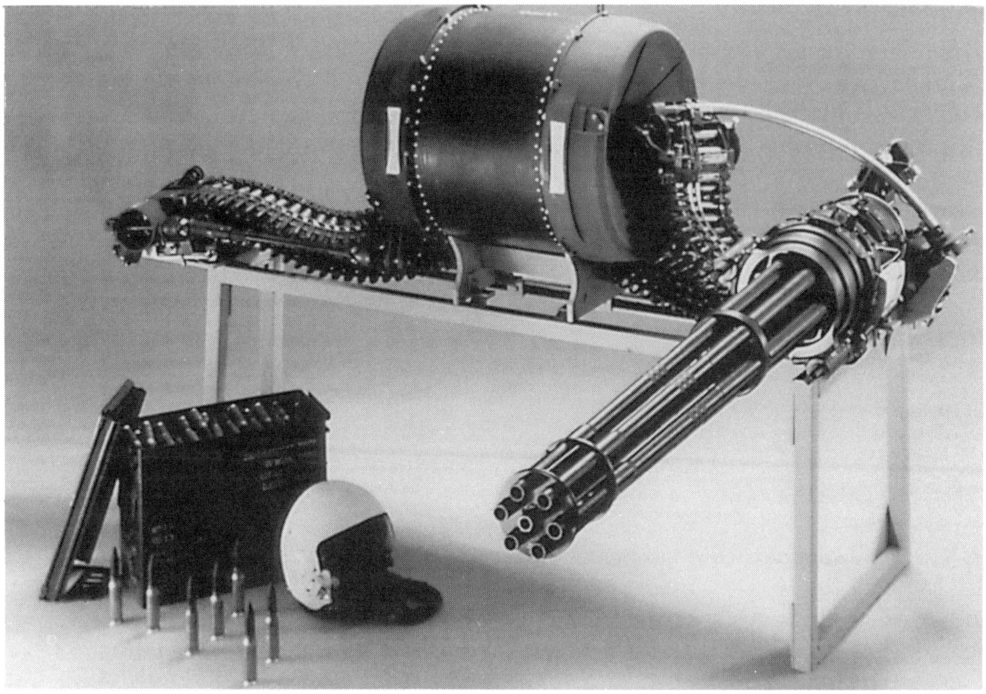

Fig. 9.2 Martin Marietta 20mm calibre Vulcan cannon fitted to the F16 (Courtesy: Martin Marietta)

A 25mm calibre Gatling cannon is mounted in the port pod whilst the starboard pod contains the 300-round, linkless feed system. A fairing between the two pods connects the gun and the feed system. The loaded system weighs 580 kg and is driven by air bled from the aircraft engines. Rate of fire of the five-barrelled cannon is 3,000 rounds per minute. The system is shown in Figure 9.1.

Internally-mounted aircraft cannon systems are usually designed to suit each individual type of aircraft. A typical example of this type of system is the Martin Marietta Gatling cannon fitted to the F16 aircraft as shown in Figure 9.2. The cannon is fitted on the port side of the aircraft and below the centre line of the nose. Ammunition is stored behind the pilot and is transferred to and from the cannon via a linkless conveyor. The weapon is a six-barrelled Vulcan Gatling cannon with a rate of fire of 6,000 rounds per minute.

The weapon on fixed-wing aircraft systems is usually fixed relative to the fuselage of the aircraft. Sighting and firing of the weapon are carried out by the pilot who must manoeuvre the aircraft to align the sights on to the target. Sights usually consist of a fixed graticule projected into the pilot's field of view in a head-up display.

Helicopter Cannon

A major recent addition to the battlefield is the attack helicopter. One of the principal weapon systems fitted to these aircraft is the medium-calibre, lightweight cannon which is used against other helicopters and ground targets. A difference between these weapons and those fitted to fixed-wing aircraft is that they usually have lower rates of fire to conserve ammunition because of the restricted payload associated with helicopters. Additionally, helicopters are relatively light aircraft so that there is usually a need to minimise peak recoil forces; this is achieved with lower rates of fire.

FIG. 9.3 Giat 20mm calibre cannon externally mounted on a Gazelle helicopter (Courtesy: Giat Industries)

FIG. 9.4 30mm McDonnell Douglas M230 chain gun mounted on the AH-64 Apache helicopter (Courtesy: McDonnell Douglas)

A variety of weapon systems have been developed for external mounting. The calibre of the cannon used for rail mounting is usually 20mm. Figure 9.3 shows an externally-mounted Giat 20mm type 20 M621, self-powered cannon fitted to a Gazelle helicopter. Rate of fire is 800 rounds per minute and the number carried is 240. Total weight of the mounting, weapon and ammunition is only 187kg. Externally-mounted weapons are fixed relative to the body of the aircraft. The sights consist of a fixed graticule, and to aim the weapon the pilot must align the aircraft with the target.

Dedicated cannon for helicopter use have, in recent years, been specifically developed for the task and may be designed to be the main armament of the aircraft. They are usually externally powered and may be single or multi-barrelled weapons. A major problem with cannon which are fixed relative to the aircraft is that they can be aimed only by the pilot, who must dedicate himself and the aircraft to that one task. To overcome this, turrets have been developed in which the cannon system is mounted so that it can move in both azimuth and elevation relative to the aircraft. This considerably increases the flexibility of the weapon system by allowing either the pilot or the co-pilot to aim and fire the cannon. One such is based on the McDonnell Douglas M230 chain gun and is fitted to the AH-64 Apache helicopter shown in Figure 9.4.

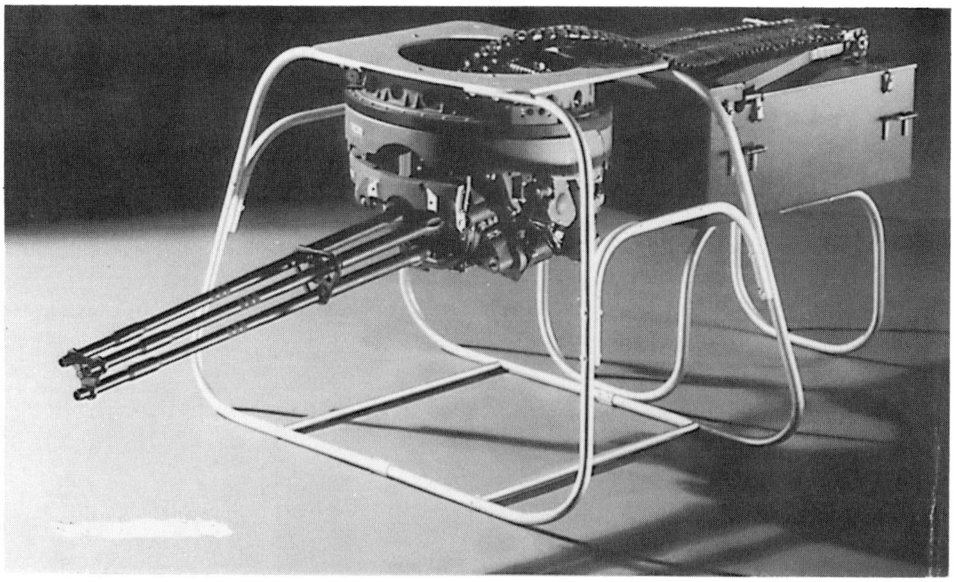

FIG. 9.5 Turret and cannon system fitted to the Cobra attack helicopter (Courtesy: Martin Marietta)

The M230 chain gun fires the 30mm calibre ADEN/DEFA round and has a rate of fire of 625 rounds per minute. The cannon is very light and weighs only 58kg.

An example of a helicopter turret using a multi-barrelled cannon is the Universal Turret System produced by Martin Marietta. This is in the US Army and Marine Corps inventory for use in their Cobra attack helicopters. The system uses the three-barrelled 20mm calibre M-197 Vulcan Gatling cannon. Figure 9.5 shows the layout of the system fitted to the Cobra helicopter.

The system consists of the turret, the cannon, the ammunition feed and the electronic controls. The drive for the turret is via two 28-volt, DC servo-driven motors, one each for azimuth and elevation movement. The 20mm cannon is capable of a rate of fire up to 2,000 rounds per minute, but different rates may be selected to suit the target being engaged. The ammunition feed uses linked ammunition and holds 750 rounds.

Slew rates for the turret are greater than 80° per second in azimuth and greater than 60° per second in elevation. Positioning of the cannon is controlled by the pilot or co-pilot through sights which are built into their helmets. A fixed graticule is projected on to the visor and the cannon is controlled to point in the direction that the firer is looking. An alternative is to use a companion missile sub-system sighting station.

Anti-aircraft/Anti-missile Systems

There are several different approaches to cannon-based anti-aircraft or anti-missile systems. Cannon used to engage aircraft or missiles use similar approaches

FIG. 9.6 Bofors Trinity ship-borne anti-aircraft system (Courtesy: Bofors Weapon System)

for increasing the hit chance against a small-fast-moving target that is exposed for only a short time. The basic approaches are to use a cannon with a relatively low rate of fire but with true proximity-fuzed ammunition, or to use a high rate of fire cannon with or without timed fuzed ammunition, or to use very high rates of fire with conventional ammunition. Whilst these approaches appear to be very different, they all use similar components in the complete system and it is the ammunition which provides the biggest differences.

An example of the low rate of fire cannon using proximity-fuzed ammunition is the Bofors Trinity anti-aircraft system. This uses the 40mm calibre L/70 cannon and the PFHE proximity-fuzed round. Rate of fire for the cannon is 330 rounds per minute. Figure 9.6 shows the ship-borne system.

For search and surveillance a radar is used. The purpose of this is to detect incoming aircraft or missiles. An integral part of the cannon system is a fire-control computer. An initial function of this is to estimate from the output of the surveillance radar the level of threat of any aircraft or missile which is detected. If there are several, the computer estimates which poses the greatest threat. Once the threat has been identified it is passed to a tracking radar. Having a separate tracking radar allows the surveillance radar to continue to search for more targets and assess their threat levels. The tracking radar gives the range and direction of the target and its crossing rate. This information is processed by the fire-control computer to calculate where the gun must be aimed relative to the target for the projectile to intercept it. The computer controls the positioning of the gun as well as carrying out all of the appropriate calculations.

The Bofors anti-aircraft cannon system is claimed to be so accurate that it tends to place all of the rounds fired in a group which is too tight to obtain the maximum hit chance. The fire-control computer can only predict where the aircraft or missile will be if it continues on the same heading the moment that the weapon is fired. If the target changes direction after the weapon is fired, the projectiles will all miss in the same place if they all follow the same path. The Trinity system therefore fires an optimum pattern distributed around the target so that one round has a high chance of hitting the target rather than all of them missing the same place. Radars are active systems that can be detected by enemy aircraft and missiles. Thus the intention and the position of the cannon system can be communicated to the target. Surveillance is therefore often carried out by a remote system

FIG. 9.7 Marconi Marksman anti-aircraft System (Courtesy: Marconi)

which passes details to the fire-control computer, which engages the tracking radar at the last possible moment to give the target the minimum possible warning of the position and intention of the cannon system. Additionally, passive tracking devices using infra-red or low-light-level television cameras may be used so that the target is unaware that it is being tracked and so does not take evasive action.

High rate of fire cannon using timed rounds are typified by the 35mm calibre KDA Oerlikon cannon firing the Oerlikon AHEAD round. An anti-aircraft system using the KDA cannon is the Marconi Marksman, shown in Figure 9.7. This uses twin cannon mounted on an armoured vehicle chassis of the customer's own choice and has a combined rate of fire of 1,200 rounds per minute. The Marksman system has all of the usual features of radar and fire-control systems except that it uses a single radar for both surveillance and tracking.

The Oerlikon AHEAD shell was described in Chapter 5. At the heart of the system is an electronic fuze which detonates the HE in front of the target to shower it with pre-formed fragments. It is vital that the fuze detonates the HE at exactly the right time for maximum effectiveness and for this the round relies upon its electronic timer. The time of flight depends upon the target range and the muzzle velocity of the shell. Target distance is given by the tracking radar,

FIG. 9.8 Goalkeeper anti-missile cannon system (Courtesy: Martin Marietta)

but muzzle velocity is subject to round-to-round variations which are affected by barrel wear and barrel temperature. The muzzle velocity of each round is therefore measured using induction coils fitted to the muzzle of the cannon. As the shell passes through the first coil it starts a timer which is stopped as it passes through a second one. The distance between the two coils is known so that the muzzle velocity can be calculated. Time of flight is calculated by the fire-control computer and is imparted to the electronic fuze as it passes through a third coil on the muzzle.

Anti-aircraft/anti-missile systems using very high rates of fire use multi-barrelled cannon such as the Meroka volley firing cannon or multi-barrelled Gatling cannon. The rates of fire for these systems may be as high as 6,000 rounds per minute. These systems also use fire control, tracking and surveillance radars. Typical of a ship-borne, anti-missile system is the 30mm calibre Martin Marietta GUA-8 seven-barrelled Gatling cannon with a rate of fire of 4,200 rounds per minute shown in Figure 9.8.

The system uses conventional kinetic energy projectiles and relies upon saturating the target area with a very large number of projectiles. These types of weapon have a fairly large dispersion ideally suited to this application to give the maximum hit chance.

Infantry Fighting Vehicle Cannon Systems

The infantry fighting vehicle (IFV) was developed from the armoured personnel carrier (APC) resulting from the widespread mechanisation of the infantry during the 1960s. The first true IFV is generally accepted as being the Soviet-built BMP-1 which was introduced into service in 1969 and which is still in use throughout the world. This vehicle has a single-man turret with a low-pressure, single-shot, 73mm calibre cannon fitted with an auto-loader and capable of firing two rounds per minute. Additionally it has a 7.62mm calibre, coaxially-mounted machine gun and a Sagger wire-guided anti-tank missile with a range of 400m. In 1981 an improved version of this vehicle was introduced. The main improvements were to the weapons system. A new two-man turret was fitted for the gunner and the vehicle commander who sat side by side separated by a new 30mm calibre, gas-operated, high velocity cannon. A second-generation, semi-automatic Spandrel anti-tank missile has replaced the obsolete first-generation Sagger. Another major improvement was the addition of two-axis stabilisation of the turret which allows the cannon to be fired accurately whilst the vehicle was moving.

In 1989 a further Soviet IFV was introduced, known as the BMP-3. This was the most advanced and sophisticated IFV of its time anywhere in the world. The two-manned turret with the gunner and vehicle commander seated side by side was retained. The main armament is a 100mm calibre, 2A70 combined gun/missile launcher and a 30mm calibre, 2A72 gas-operated cannon. The 100mm calibre gun fires both laser-guided missiles and conventional HE rounds. The missiles are intended for use against armoured targets at ranges up to 4,000m. It has a shaped charge warhead which can penetrate approximately 500mm of rolled homogeneous (armour (RHA).

The 30mm calibre 2A72 cannon can be fired single-shot or in bursts and has a dual belt-feed mechanism to enable one of two types of ammunition, either HE or kinetic energy, to be fired. The rate of fire is 500 rounds per minute. A coaxial 7.62mm calibre machine gun is also fitted that has a rate of fire of 250 rounds per minute.

An auto-loader is used with the 100mm calibre gun for both missile loading and loading conventional ammunition. Up to six missiles are carried and 40 rounds of 100mm calibre HE ammunition. The rate of fire using the auto-loader is up to 10 rounds per minute. Maximum elevation and depression is +60° and –5°. The reason for such a high elevation is probably more to allow the weapons to be used in mountainous country rather than to give the vehicle an anti-helicopter capability.

An important feature of the cannon system is that it is fitted with a fire-control computer. This, and the ability to fire on the move, by means of turret stabilisation, makes for a very flexible weapons system capable of high hit chances under a wide variety of circumstances. The fire-control system also reduces engagement times. The gunner places his sights on the target, tracks it for a short distance, and the computer, in combination with a laser rangefinder, estimates the range, the fall in projectile with range, the amount of target lead required, and will point the gun in the correct direction to hit the target. Additionally, the computer will also take into account the elevation of the target and the sideways angle of the vehicle if it is not on flat ground.

An additional advantage of a fire-control system is that targets at longer ranges may be engaged and the first round (and subsequent rounds) hit chance will be increased. This will reduce the number of required stowed rounds, which is a major consideration for any vehicle where storage space is always severely restricted.

Sights consist of a × 8 day sight and a × 5.5 night sight for the gunner who also has a × 2.6 secondary day sight. The commander has a dual day/night sight of × 1.2 and × 4.

The BMP-3 almost certainly points the way ahead for weapons control but not for the number of weapons squeezed into the turret. A 100mm calibre weapon in addition to a 30mm calibre cannon and a coaxial machine gun reduces the turret space to a level that would probably be unacceptable to Western nations. Also, the work load for the vehicle commander would be very demanding with such a wide variety of weapon systems to be controlled.

Western IFVs are surprisingly conservative in the design of their weapons systems. Many are being introduced at the present time that do not have turret stabilisation so that they cannot engage targets while on the move. There seems to be a reluctance to fit gun stabilisation and there appear to be no plans to fit a fire-control computer, apparently because of the costs involved. Most of the effort in the West appears to be directed towards an increase in the target effect of the main armament by increasing the calibre of the cannon. This will also increase the effective range of the weapon, providing that the target can be hit sufficiently often.

Coaxial Cannon for Main Battle Tanks

One of the first coaxial cannons fitted to a main battle tank (MBT) was the 20mm Polson cannon fitted to the Centurion tank of the Second World War. This was not a very effective or popular weapon because, amongst other problems, there was the danger of toxic gas build-up in the turret caused by the Polson's advanced primer ignition (API) blow-back action. Currently, one of the few modern MBTs fitted with a coaxial cannon is the French AMX-30. This has a 20mm calibre Giat cannon; the usual 7.62mm calibre coaxial machine gun is also retained.

The usual reason for fitting a coaxial cannon is to enable targets to be engaged that could not be defeated by the 7.62mm calibre coaxial machine gun but where the use of the main tank gun would be a gross over-kill. If there were sufficient targets in this category there could be a space saving because of the need to carry fewer rounds for the main armament. Few nations at the moment think that this is the case, but should the proposed future 140mm calibre tank gun be introduced then this may change. The 140mm calibre ammunition will be so large that it will almost certainly require an auto-loader, which may eliminate the need for a human loader so that there will be more room in the turret for a cannon, its feed system and its ammunition. Most nations undertaking design studies for 140mm tank gun systems are also considering the use of a coaxial cannon. An additional advantage of a tank cannon is that it would be capable of firing ammunition with similar ballistic characteristics to the main armament. This would allow it to be used as a ranging system in the event of a fire-control computer failure.

10.

Future Cannon Developments

The Need for Improvements

Targets are constantly improving their armoured protection. In addition, they are increasingly more agile and difficult to hit so there is a need for greater armour-penetration performance and reduced projectile flight times to increase hit probability. Because the performance of a cannon is directly related to the calibre of the weapon there is a trend for cannon calibres to increase. However, because cannon systems are almost always vehicle mounted, the space for accommodating the weapon and its ammunition is severely restricted. A new generation of weapons is being developed based upon new ammunition configurations that make lighter and more compact designs possible. External power is almost exclusively used for this new generation, which allows important weight and space savings to be made.

The volume in IFV turrets is highly restricted so that the space occupied by the cannon system must be minimised. However, there is considerable latitude when the mounting position of the cannon within the turret is decided upon. It is highly desirable that the cannon be balanced at its trunnions otherwise balancing gear is required, which adds to weight, cost and complexity. But the most important factor is the inboard length of the cannon. A minimum inboard length gives the crew the maximum room to load and service the weapon, but the further forward the gun is mounted then the greater will be the reduction in elevation and depression, because movement of the weapon will be restricted by the narrowing of the turret roof and the turret roller race assembly. Raising the height of the turret to alleviate this problem has an adverse effect upon the vehicle silhouette. Thus there is a limit to how far forward the weapon can be mounted. To reduce this problem it is necessary for the length of the cannon behind the trunnions to be kept to a minimum. Ammunition length has a direct bearing upon the length of the cannon behind the trunnions so it is an advantage to have short ammunition. Additionally, as the calibre of the ammunition increases, there is also an increase in the size of the feed system. The depth of the feed system is proportional to the length of the ammunition. Ammunition stowage space is extremely restricted for all vehicle-mounted weapons. Increasing the calibre of a weapon system increases the volume of the ammunition and the stowage space that it needs. Because space is limited, an increase in ammunition calibre results in a reduction in the amount of ammunition that can be carried.

35/50mm Ammunition System

The original 35/50mm calibre ammunition system concept was developed as part of the German Army's programme to upgrade the Marder IFV. Although the upgrade of the vehicle has been cancelled, Rheinmetall have continued with the development of their cannon, the Rh503, and the 35/50mm ammunition that it fires. The 35/50mm calibre ammunition system consists of a standard 35mm calibre × 228mm NATO round and a 50mm calibre round that fits into the same space envelope. The 50mm calibre round is designed with a parallel cartridge case, and with the base of the cartridge case and the length of the round the same as for the 35mm calibre ammunition. Thus a weapon can be designed to fire the 35mm calibre round and, when required, the weapon may be increased in calibre by changing the barrel. A penalty is paid in limiting the length of the 50mm calibre round to that of the 35mm calibre round in that the muzzle energy is restricted: the muzzle energy lies between that expected of a conventional bottle-shaped 40mm and a 45mm calibre round.

Figure 10.1 shows the Rheinmetall cannon with the 35mm and 50mm calibre barrels.

Because 35mm ammunition is based upon the 35mm calibre × 228mm round for which there is a NATO Standardisation Agreement (Stanag), a wide range of ammunition has been developed for that calibre. Altogether there is therefore little technical risk associated with ammunition development; so all of the initial development of the cannon will be directed towards using the 35mm calibre round. Development of the 50mm calibre round will not be undertaken in earnest until there is a more pressing need for the larger calibre, but the configuration is fairly conventional and little technical risk will be involved. Figure 10.2 shows the 35mm and the 50mm calibre APFSDS-T round.

An additional advantage of the 35/50mm concept is that if the larger calibre were adopted the smaller one would still be available for training purposes using

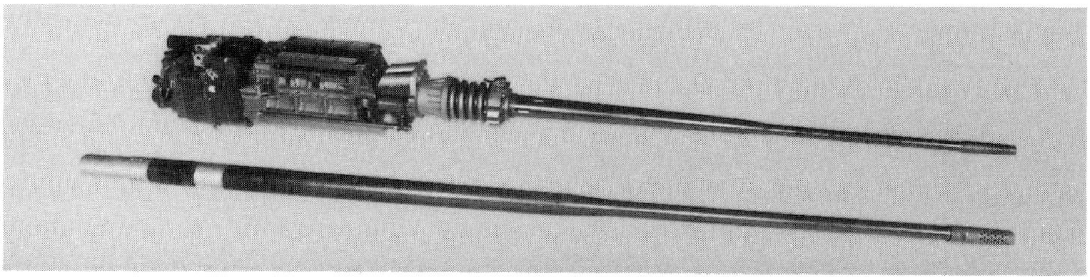

FIG. 10.1 Rheinmetall Rh503 cannon with 35mm and 50mm calibre barrels (Courtesy: Rheinmetall GmbH)

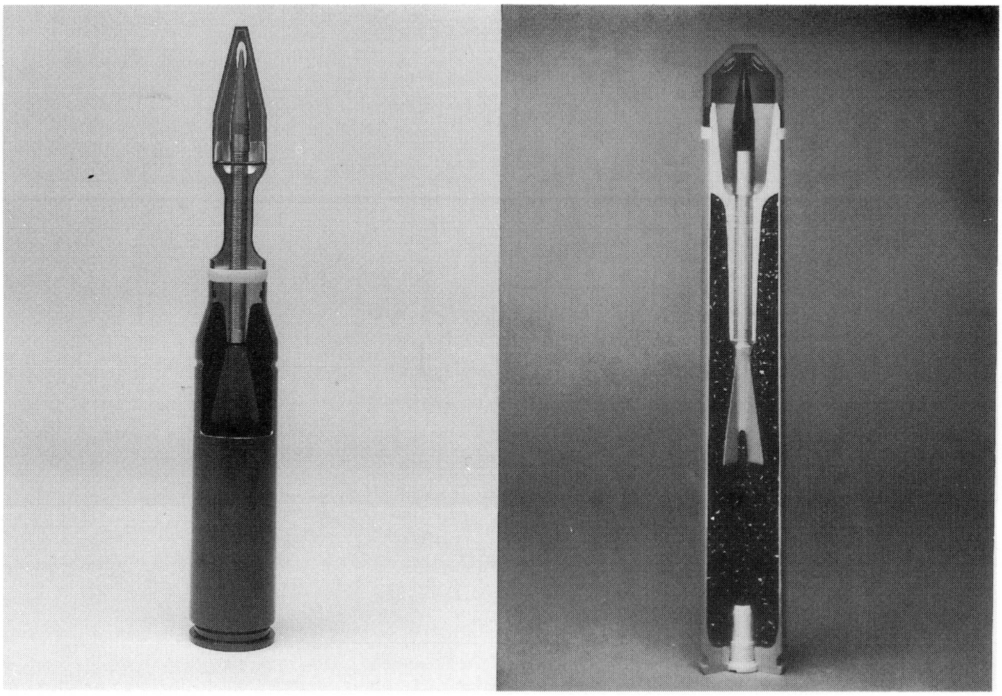

FIG. 10.2 35mm and 50mm calibre APFSDS-T rounds (Courtesy: Rheinmetall GmbH)

lower cost ammunition. The 35/50mm calibre concept is an innovative idea for IFVs. It enables a conventional system to be designed based upon only a modest increase in calibre that would meet the medium-term armour-penetration requirement. Should the long-term need be for an even greater performance, there would be little cost involved in changing to the 50mm calibre. However, a disadvantage with this approach is that the original system developmental costs would be higher because of the need to design it at the outset to feed and fire the two differently sized rounds. It is the feed system where special consideration must be given. Control of the round must be via the base of the cartridge case because that, and its length, are the only points of commonality that the two rounds have. An additional burden is that the 35mm calibre system must be designed to withstand the firing stresses of the 50mm calibre ammunition, so that the weapon will be more massive than if designed for the smaller ammunition alone.

Another area for special consideration is the recoil system. Recoil forces will be considerably higher for the larger calibre round and the design must reflect this.

Chain Gun Family of Weapons

McDonnell Douglas are in the process of enlarging their range of weapons which will be based on their externally-powered cannon, which uses an endless

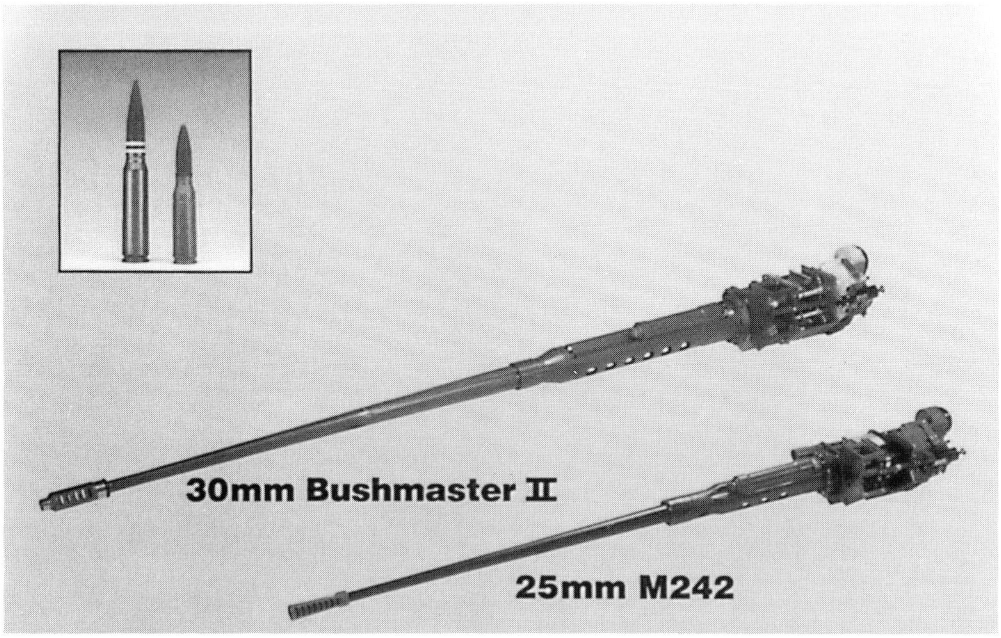

FIG. 10.3 25mm calibre Bushmaster and 30mm calibre Bushmaster II cannon (Courtesy: McDonnell Douglas)

chain to control the operating cycle. The 25mm calibre M242 Bushmaster fitted to the Bradley M2/M3 IFVs has already been described. A Bushmaster II based upon the original M242 has been developed, which fires the 30mm calibre GAU-8 round that is similar in size and performance to the 30mm calibre RARDEN cannon round. Figure 10.3 shows the original Bushmaster and the larger Bushmaster II cannon.

A Bushmaster III has also been designed that will fire the 35mm × 228mm NATO round. This has been designed to take advantage of the 35/50mm concept. Work is also proceeding with a Bushmaster IV that will be capable of firing the 40mm calibre Bofors L/70 round.

The chain gun family of weapons has the advantage of being light and compact and can take full advantage of the conventional ammunition being developed. It has a mechanically simple construction and has proved to be highly reliable. However, as the length of the round increases then so does the length of the cannon and the depth of the feed system that it employs. It is inevitable that the size of the cannon system will rise substantially as the calibre increases. Whilst the 35/50mm concept significantly helps in restraining the increase in ammunition length, a more innovative approach is needed in order to have a real impact on the problem.

Cased Telescoped Ammunition

Cased telescoped ammunition (CTA) is a concept that was developed a number of years ago. It is only recently that there has been any real incentive to exploit the idea. Contemporary IFVs with their big increase in armour protection, such as the Soviet BMP-3, have focused the need for cannon of greater calibres and the requirement for a different design approach to allow them to be fitted into current and future vehicles. CTA has the potential for greatly reducing the size and weight of these larger calibre systems. A CTA projectile is completely contained within the cartridge case and is surrounded by the propellant. The cartridge case is made larger in diameter to accommodate this and the shape is a cylinder with parallel sides. This considerably reduces the length of the round and makes more efficient use of the space that it occupies. Currently the United States, France and Britain are developing a 45mm calibre CTA round and the cannon that fires it for possible installation in their current IFVs and for their future vehicles. Figure 10.4 shows sections through typical APFSDS, APDS and HE rounds.

The United States has been pursuing the CTA concept under its COMVAT programme, which was a study for upgrading the Bradley IFV. Both prototype ammunition and cannon have been developed. Giat of France have been at the forefront of CTA technology for many years and have also developed prototype ammunition and cannon. Britain has been late in showing an interest in this

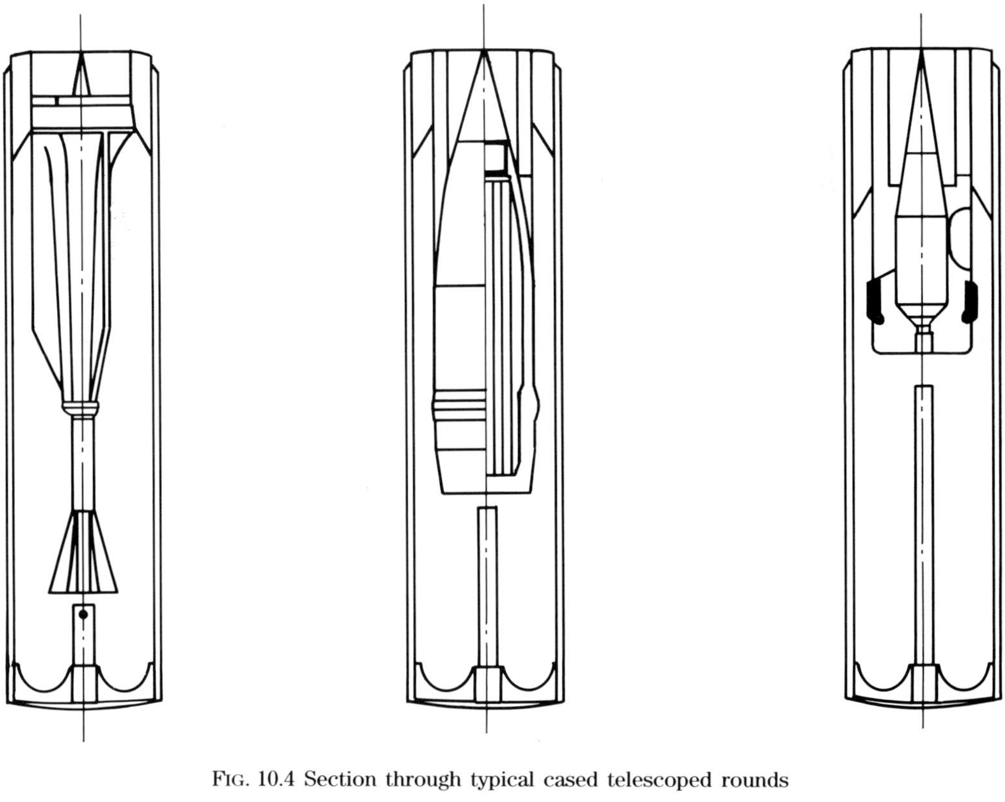

FIG. 10.4 Section through typical cased telescoped rounds

TABLE 10.1
Number of rounds that can be carried in the same storage space in the Bradley M2 and the Warrior IFV.

Bradley M2					
Calibre (mm)	25	30	35/50	40	45 CTA
Rounds	900	550	245	160	195
Warrior					
Calibre (mm)	30	35/50	40	45 CTA	
Rounds	200	90	60	70	

TABLE 10.2
Overall length of a complete round of ammunition for different calibres.

Calibre (mm)	25	30	35/50	40	45 CTA
Overall length (mm)	223	284	390	530	305

type of technology but is catching up fast. The Defence Research Agency at Fort Halstead has entered into a joint technology demonstrator programme with British Aerospace's Royal Ordnance Division, which started in April 1993. They have already fired prototype ammunition and their cannon studies are well advanced. An important factor in the development of this new ammunition concept is that a draft NATO Stanag has already been produced to ensure that the ammunition developed by each nation will operate in the others' weapons.

The usual advantage stated for CTA is the reduction in ammunition volume compared with bottle-shaped ammunition so that more rounds can be carried for the same calibre. This certainly is a major consideration. Table 10.1 shows the number of rounds that are currently carried in the Bradley M2 and in Warrior, and the number of larger calibre rounds that can be carried in the same space. Increasing the calibre has a dramatic effect on the number of rounds that can be carried. It should be noted that the 35/50mm system allows the greatest number of rounds to be carried for those calibres big enough to meet the future threat in armour protection.

Of equal importance is the length of the round. This influences the inboard length of the cannon and the depth of the feed system. The overall length of different calibre rounds is given in Table 10.2. This shows the very large reduction in ammunition length achieved by CTA technology.

The parallel-sided case of the CTA round has an additional advantage of allowing preformed, consolidated propellant to be used. This can be preformed into a cylinder and loaded through the neck of the case. Solid propellant gives an increase in the weight of the propellant that can be loaded into the case, compared with granular propellant, of about 30 per cent. The ammunition being developed in

FIG. 10.5 45mm calibre CTA COMVAT cannon (Courtesy: Allient Tech Systems)

the United States uses consolidated propellant but France and Britain are currently using standard granular propellants.

One of the most important features of CTA is that it allows a complete rethinking of the cannon system design. Almost all of the CTA cannon designs being investigated have a fixed breech and rotating chamber. The operating system may be considerably simplified by using push-through ejection, which is made possible by the parallel-sided case of the CTA round. Typical of the approach is the American-designed cannon as part of the COMVAT programme, as shown in Figure 10.5.

The operating cycle for the COMVAT cannon is shown in Figures 10.6a to 10.6f.

The base of the case has a groove into which a claw fits to position the case axially in the chamber. It is also required to provide a means of removing live rounds from the chamber. There is a clearance between the front of the chamber and the barrel. Both ends of the case are spherical in shape: the radius is that of the rotating chamber. The rate of fire for the COMVAT cannon is 200 rounds per minute. The French Giat cannon is capable of a rate of 400 rounds per minute: it can achieve this by rotating the breech through a smaller angle, which considerably speeds up the operating cycle.

The American COMVAT system feeds ammunition through the hollow left-hand trunnion of the cannon and ejection takes place through the hollow right-hand trunnion. This ensures that during the elevation and depression of the gun there is no relative movement between the ammunition feed and the chamber.

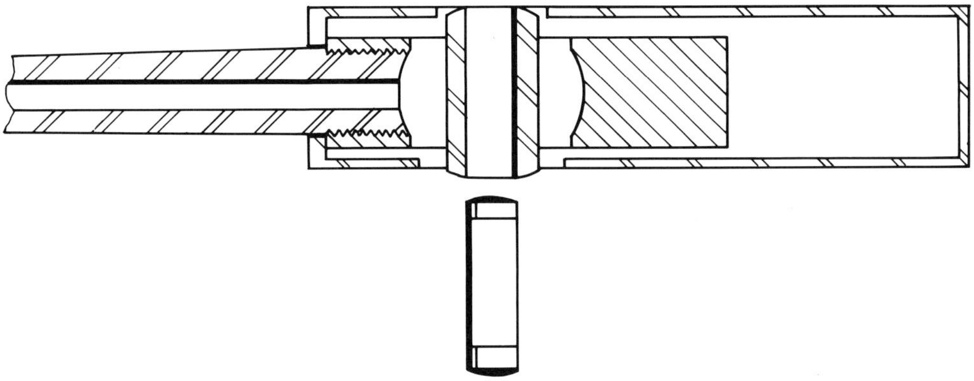

Fig. 10.6a For loading the chamber rotates through 90° to the bore

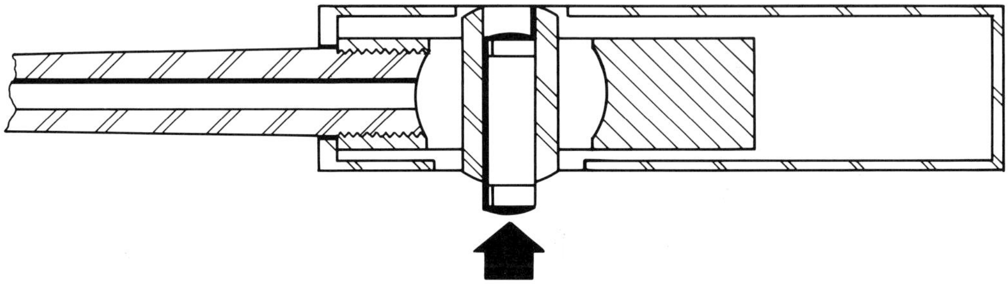

Fig. 10.6b A round is rammed

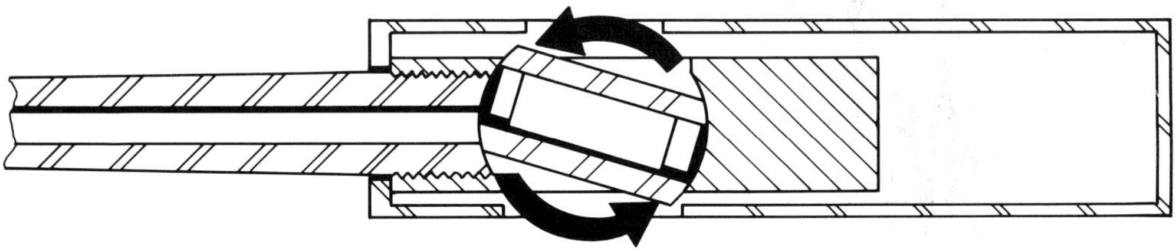

Fig. 10.6c The chamber rotates to align with the bore

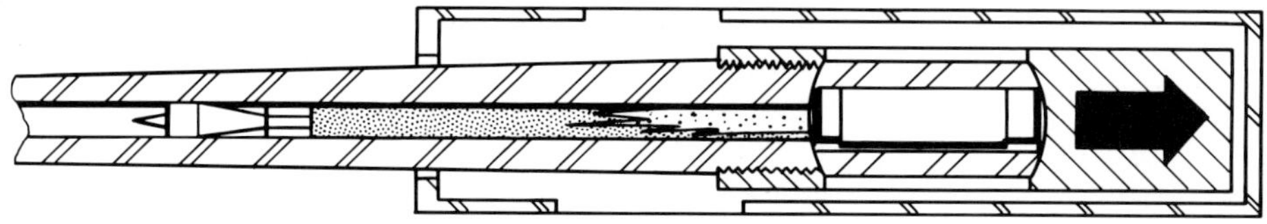

FIG. 10.6d The round is fired and the gun recoils within the box receiver. Energy is stored in a spring system which returns the chamber to the load position

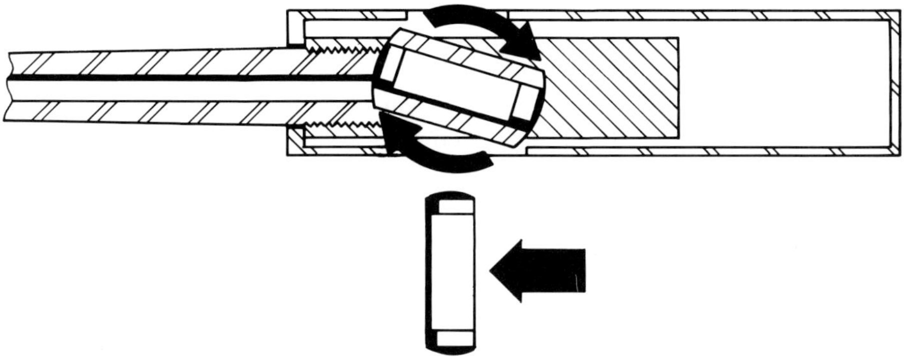

FIG. 10.6e The chamber is rotated through 90°

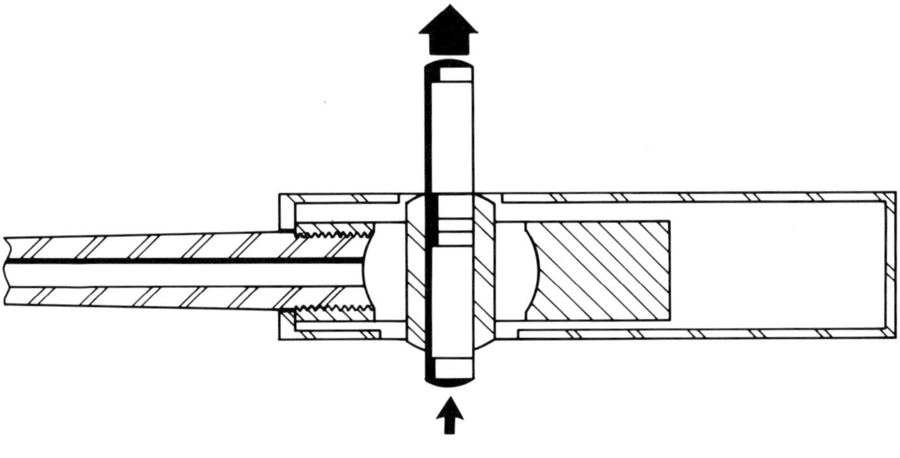

FIG. 10.6f A fresh round is rammed pushing the spent case out of the forward end of the chamber

An additional advantage of a rotating chamber system is that ammunition feed can be at a right angle to the axis of the gun. Thus the protrusion of the feed system into the turret is considerably reduced. Whilst the COMVAT design of ammunition feed has many advantages, there are other approaches which are being considered by both Britain and France. Any final solution will much depend upon the actual circumstances of a particular application. Two examples being studied by Giat are shown in Figure 10.7. The first shows a system similar to that being developed for COMVAT but with the addition of provision for turning the ammunition before feeding to align with the chamber which rotates through the same angle. The second example shows the same arrangement but turned on its side. Ammunition feed is reduced still further in the amount of space needed to accommodate it.

While Britain and France continue work on CTA technology aimed primarily at IFV applications, the focus of American interest has switched to aircraft and anti-aircraft/anti-missile use. Martin Marietta have developed a five-barrelled Gatling cannon which fires cased telescoped ammunition. This weapon has been

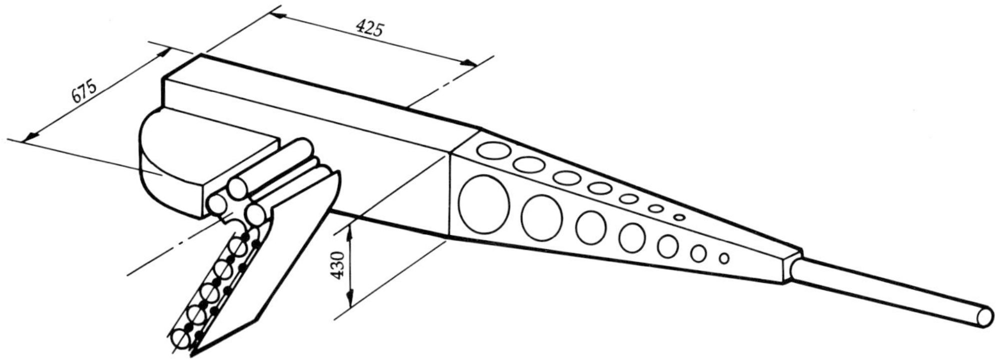

Horizontal Feed Arrangement

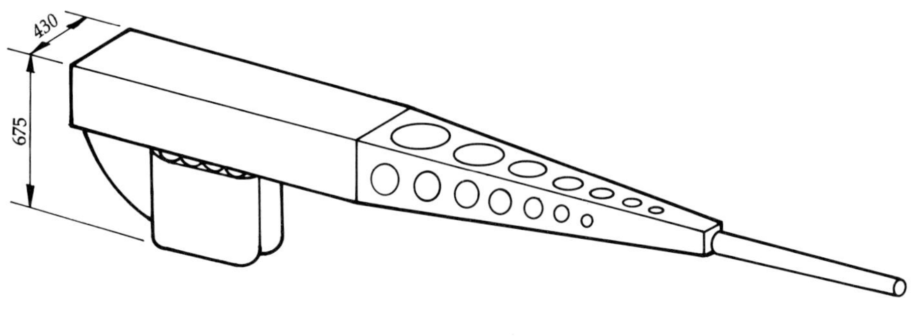

Vertical Feed Arrangement

FIG. 10.7 Examples of different feed systems being investigated by Giat (Courtesy: Giat Industries)

FIG. 10.8 20mm calibre CTA round and conventional 20-mm round
(Courtesy: Martin Marietta)

designed with half the number of total parts and one-third the number of moving parts compared with the original 20mm Vulcan Gatling cannon. Its rate of fire is 6,000 rounds per minute. Muzzle velocity achieved by the 20mm CTA round is 1525m/s. This very high velocity gives almost twice the muzzle energy compared with conventional 20mm ammunition. More important, however, is the large reduction in flight time of the projectile from the muzzle to the target, which considerably increases hit probability. Figure 10.8 shows the 20mm calibre CTA round compared with a conventional 20mm round.

Conclusions

Almost all cannon developments are being directed towards IFV applications. The driving force behind these is the considerable increase in contemporary IFV armour protection and predicted improvements. To defeat these higher levels of protection will require considerable improvements in weapons performance over the current generation of weapons. This can only be achieved by increasing weapon calibre. To fit these weapons into current vehicles will need a radical approach because the currently-used designs made bigger cannot possibly fit into the space available. The 35/50mm approach, using whatever cannon, is the safest option so far as technical risk is concerned but would require so much space that the effectiveness of the system may be significantly compromised. CTA technology offers the greatest potential but has the greatest associated technical risk.

Glossary

Accelerator
device used with short-recoil operating system to accelerate the breech block to the rear of the receiver

Adiabatic
no heat flow to or from the system

Adiabatic flame temperature
the maximum temperature achievable under ideal conditions

Advanced primer ignition
blow-back operating cycle in which the primer is initiated before the cartridge is fully chambered

AFV
armoured fighting vehicle

Aim off
The angular displacement of a gun in azimuth required to hit a moving target

Air burst
an explosion of a bomb or projectile above the surface

All burnt
the stage in the sequence of events when a gun is fired at which the propellant is completely consumed

AP
armour-piercing

APC
armoured personnel carrier

APDS
armour-piercing discarding sabot: ammunition where a sub-calibre penetrator of heavy metal is surrounded by a sabot which supports the penetrator in the barrel. Thus the largest possible base area is presented to the propellant gases over which the pressure can be exerted. The sabot separates from the penetrator when it emerges from the barrel, leaving the penetrator to continue to the target

APFSDS
armour-piercing fin stabilised discarding sabot: ammunition in which a sub-calibre fin-stabilised projectile is fired from a gun of larger calibre by means of a sabot which is discarded at the muzzle

APHC
armour-piercing hard core: ammunition with a heavy metal hard core to enhance armour penetration

Azimuth
the horizontal plane

Barrel condemnation criteria
criteria used for determining when a barrel is no longer serviceable

Base fuze
fuze located in the base of a projectile

Belted case
cartridge case reinforced at the base with an additional band of case material on the outer surface

Blow back
weapon operating system which uses the force of the expanding propellant gases acting against the base of the bolt to cycle the weapon, the bolt not being locked to the barrel

Bolt
part of the breech mechanism of a gun which slides backwards and forewords to open and close the breech

Calibre
land-to-land diameter of the gun barrel

Cannon
Direct fire weapon between 20mm and 50mm calibre

Cartridge
a complete round of fixed or semi-fixed ammunition

Chain gun
externally-powered gun manufactured by McDonnell Douglas using a chain to control the operating cycle

Chamber
the smooth section at the breech end of the ordnance shaped to accommodate the cartridge

Chamber pressure
the pressure generated in the chamber by the burning propellant

Choking
the condition of maximum mass flow which may pass through a nozzle for given upstream conditions

Closed bolt
operating system where the weapon has a round in the chamber when the trigger is pulled to fire the weapon

Commencement of rifling
the point at the breech end of the barrel where the rifling reaches maximum depth

Consistency
the measure of the degree of spread of a group of rounds about their mean point of impact

Cook-off
the initiation of a round in the chamber of a gun by the conduction of heat from the chamber of the weapon

Course correcting ammunition
ammunition with small explosive charges around their circumference that can deflect the projectile towards the target

Cradle
the component of the superstructure that carries the ordnance and allows it to slide axially on recoil

CTA
cased telescoped ammunition

Delayed blow back
blow-back operating system with a device for slowing the opening movement of the breech block

Double-based propellant
propellant consisting of a mixture of nitrocellulose and nitroglycerine

Drag coefficient
in ballistics, the ratio of drag pressure to dynamic pressure

Driving band
a band fitted round a projectile for the purpose of engaging the rifling

Externally-powered weapons
weapons which do not use the power generated by the propellant gas to cycle them

Fire control system
those elements of the weapon system used to calculate the correct fire position of the gun

Firing pin
a percussion device used to initiate a stab sensitive detonator

Fuze/fuse
a device used to detect the target and allow the detonator to be activated

Gas operated
method of self-powering a gun by tapping of high pressure propellant gas from the barrel.

Gatling gun
a gun with a cluster of rotating barrels for sustained high rates of fire

GPMG
general purpose machine gun

Graticule
sighting or reference mark etched on a glass plate and inserted into the field of view by placing it on an image plane

Hammer forging
method of producing the rifling grooves in a barrel by reducing the diameter of the barrel on to a mandrel

Hang fire
an undesirable delay in the functioning of a firing system

Hard extraction
the failure in or difficulty of extracting a fire case from the chamber of a weapon due to excessive friction between the case and the chamber walls

HE
high explosive

HEAT
high explosive anti-tank; shaped charge warhead

HESH
high explosive squash head: ammunition in which the HE charge is designed to deform into a pad against a hard target before detonating to propagate shock waves through the target material

HMG
heavy machine gun

IFV
infantry fighting vehicle

Jacket
the outer covering of a bullet

Lands
the raised portions between the grooves of the rifling in the barrel

Locked blow back
operating cycle of a gun where the breech block is locked to the barrel for the first part of the cycle and then unlocked and relies on blow-back action to cycle the weapon

Long recoil
operating cycle of a gun using the recoil energy and in which the barrel and breech, locked together, recoil the length of an unfired round

Mach number
the speed of a projectile divided by the ambient speed of sound

Machine gun
automatic gun of less than 20mm calibre capable of sustained rapid rates of fire

Magazine
a store or container for ammunition or explosives

MBT
main battle tank

Mean point of impact
the centre point of the spread of rounds fired from a gun at the same line and elevation

Mil
an angular measurement, equal to 1/6,400 of a circumference

Multi-purpose ammunition
projectiles with electronic fuzes that can be programmed to perform a wide range of functions

Muzzle brake
a device attached to the muzzle of a gun to deflect muzzle gases rearwards to reduce the contribution of the propellant gases to the recoil momentum of the weapon system

Muzzle energy
the kinetic energy of a projectile at the muzzle

Nitride
surface treatment of special steels to produce a hard, wear-resisting layer

Nitrocellulose
an explosive produced by the nitration of cellulose from cotton or wood pulp

Nitroglycerine
glyceryl 1,2,3-trinitrate, an explosive produced by the nitration of glycerine

Nitroguanidine
a secondary high explosive used as a major ingredient of triple-based propellants

Obturation
sealing of the breech to prevent the leaking of propellant gases

Open bolt
operating system where the breech block is held to the rear and there is no cartridge in the chamber of the gun when the trigger is pressed to fire the weapon

Ordnance
the group of components in a gun, comprising the barrel and its attachments, the breech and the firing mechanism

Primer
cap in the base of the cartridge case with an initiation compound for igniting the main propellant charge

Progressive (gain) twist rifling
rifling form where the twist rate increases from breech to muzzle

Proximity fuze
a fuze that is initiated by the close proximity of the target

Pure blow back
operating system which relies on the momentum of the breech block to constrain the cartridge case in the chamber when the weapon is fired

Pyrotechnic
a material capable of combustion when correctly initiated to provide a special effect

Receiver
main body of the gun in which the breech block moves

Recoil
the rearward movement of the ordnance on firing

Recouperator
the component of the recoil mechanism that returns the recoiling parts to their original position after firing

Revolver cannon
gas operated aircraft cannon using a number of revolving chambers to enable very high rates of fire to be achieved

RHA
rolled homogeneous armour

Rifling
helical grooves cut into the bore of a barrel to impart spin to the projectile

Rim
raised edge of the base of the cartridge case for the purpose of extracting the fired or unfired case from the chamber of the gun
Rimless
a cartridge case which is the same diameter as the rim
Rotating element
externally-powered gun using a rotating element to convert circular movement into reciprocating movement
Round
a complete assembly of a projectile, propellant charge in a cartridge case, and means of ignition
RRTR
reduced range training round: ammunition with a projectile that ballistically matches the service round over short ranges but with a short maximum range
Sabot
lightweight carrier in which a sub-calibre projectile is centred to allow the firing of the projectile from a larger calibre gun
Salvo-firing cannon
cannon with a number of fixed barrels which can be fired together or in groups
SAPHE
semi-armour piercing high explosive
Self-powered cannon
cannon which use the energy of the propellant to cycle the weapon
Shell
a hollow projectile, filled with high explosive or other material, fired from a gun
Single-based propellant
a propellant consisting mainly of nitrocellulose
Short recoil
operating system of a gun using the recoil energy in which the barrel and breech, locked together, recoil less than one-quarter the length of an unfired round before unlocking and the barrel comes to rest with the breech block moving to the rear
Sliding wedge
breech locking method where a wedge-shaped plate moves at right angles to the barrel axis and locks into cut outs in the breech face
Stanag
NATO Standardisation Agreement
Tracer
pyrotechnic fitted to the base of a projectile which enables the firer to observe the trajectory and point of impact
Trajectory
the curve described by the centre of gravity of a projectile during its flight
Trunnions
the projections about which a gun elevates
USTOL
vertical short take off and landing

Warhead
that part of a projectile containing materials intended to inflict damage

Wear
the removal of material from one surface by the mechanical action of another upon it

Yaw
the angle between the axis of the projectile and the tangent to the trajectory

Young's modulus
ratio of stress to strain for a material over its elastic region

Index

ADEN/DEFA 2, 57, 67
Adiabatic flame temperature 74
Advanced primer ignition 5
AHEAD-MUNITION 61, 103
Air burst 64
Air craft cannon systems 96
All burnt 41
Ammunition handling 89
Angle of twist 36
Angular acceleration 37, 38
Annular ring spring 80
Anti-aircraft/missile systems 60, 100
Apache helicopter 31, 99
APDS 42, 52, 56, 58, 60, 76, 111
APFDS 53, 56, 76, 111
APHC 53, 56
Armour penetration 56

Balloting 20
Barrel condemnation criteria 43
Barrel design 36
Barrel heating 41
Barrel length 36, 41
Barrel life 36
Barrel manufacture 39
Barrel materials 40
Behind armour effects 56
Belt feed 92
BMP 104, 111
Blow back cannon 5, 7
Bofors 3, 11, 42, 54, 63, 64, 67, 88, 107
Bore temperature 42
Bradley 30, 111
British Aerospace 63

Cartridge case 68
Carousel 90
Cased telescoped ammunition 111, 116
Case rim 68
Chain gun 29, 92, 99, 109
Chamber pressure 38, 41
Chromium plating 41
Coaxial cannon 106
Cobra helicopter 100
Cook-off 14, 21, 67
COMVAT 113
Course correcting ammunition 63

Delayed blow back 5
Detatchable magazine 90
Dispersion 44, 85, 86
Drag coefficient 59
Doppler fuze 61
Driving band 36, 37, 38, 44, 49, 53, 75
Dual feed 30, 34

Electrical primers 7, 73
Electrochemical machining 40
Electronic fuze 47, 64
Engraving force 36
External power 22
Extraction force 71

Fatigue 44
Fin stabilised 36, 53, 75
Fixed breech 21
Fixed zero mount 84
Flight times 54
Floating mount 85
Fluted chamber 72
Frangible round 61
Friction brake units 81

Gas operated cannon 16, 72
Giat 99, 113
Gas recouperators 81
Gatling cannon 23, 72, 94, 96, 104

Groove depth 36
Groove width 36
Gyroscopic stability 36, 44, 75

Hang fire 22, 24, 30, 33, 34
Helicopter cannon 98
HESH 57
High explosive shells 50
Hit chance 54, 60, 97, 104, 105
Hydraulic units 82

Impact fuze 1, 48, 64
IFV cannon systems 104
Incendary 50

Kill probability 51, 54, 61

Land width 36
Linkless conveyors 94
Long recoil 8

Marconi Marksman 103
Marder 33, 108
Martin Marietta 23, 90
Mauser 3
Maximum effective range 54
McDonnell Douglas 29
Mechanical fuze 47
Meroka 27, 104
Multi-purpose ammunition 64
Muzzle break 85

Nitriding 40

Oerlikon 2, 3, 47, 53, 61, 62, 88

Oto Melara 63
On-gun magazines 88

Phosphating 72
Practice ammunition 50, 58
Primers 73
Progressive twist 38
Propellants 74
Proximity fuze 62
Pure blow back 5

RARDEN 2, 10, 41, 42, 58, 79, 91
Recoil operation 8
Recoil distance 78
Recoil energy 78
Recoil force 78
Recoil units 77
Recoil velocity 78
Reduced range training round 58
Revolver cannon 19, 37, 96
RHA 55
Rheinmettall 3, 7, 33, 108
Rifling form 36
Rigid mounts 78
Rotating bolt 17, 24, 29, 33
Rotating element cannon 33
Royal Ordnance 58, 61, 112

Sabot 52, 53, 76
Safety trace 58
Salvo firing cannon 27
SAPHE 47, 51, 56
Self destruct 47, 54, 64
Self powered cannon 4, 21
Set up 43
Short recoil 8, 12
Sliding wedge 10
Spring recoil units 79

Target array 1, 2, 47
Toxic gases 10, 18, 71, 88
Trajectory diagrams 54
Turret stabilization 105
Twist rate 36, 37

Vulcan cannon 3, 98, 100, 117

Warrior IFV 112
Water cooling 42